Fiber Optics - Technology and Applications

Edited by Guillermo Huerta-Cuellar

Published in London, United Kingdom

IntechOpen

Supporting open minds since 2005

Fiber Optics - Technology and Applications
http://dx.doi.org/10.5772/intechopen.94790
Edited by Guillermo Huerta-Cuellar

Contributors
Lokendra Singh, Nitesh Kumar Agarwal, Himanshu Bartwal, Bhupal Arya, Taresh Singh, Muhammad Irfan Anis, Hamdan Ali, Sunil Sharma, Lokesh Tharani, Alaaeddine Rjeb, Habib Fathallah, Mohsen Machhout, Ulrich H.P H.P. Fischer, Zhipei Li, Dong Guo, Ran Gao, Alexander Bortsov, Hongbao Xin, Xing Li, Amado M. Velázquez-Benítez, Juan Hernández-Cordero, Oscar González-Cortez, Rodolfo A. Carrillo-Betancourt, Matthias Haupt, Peter Kußmann

First published in London, United Kingdom, 2021 by IntechOpen
IntechOpen is the global imprint of INTECHOPEN LIMITED, registered in England and Wales, registration number: 11086078, 5 Princes Gate Court, London, SW7 2QJ, United Kingdom
Printed in Croatia

British Library Cataloguing-in-Publication Data
A catalogue record for this book is available from the British Library

Additional hard and PDF copies can be obtained from orders@intechopen.com

Fiber Optics - Technology and Applications
Edited by Guillermo Huerta-Cuellar
p. cm.
Print ISBN 978-1-83969-626-8
Online ISBN 978-1-83969-627-5
eBook (PDF) ISBN 978-1-83969-628-2

For EU product safety concerns:
IN TECH d.o.o., Prolaz Marije Krucifikse Kozulić 3, 51000 Rijeka, Croatia, info@intechopen.com or visit our website at intechopen.com.

Meet the editor

Guillermo Huerta Cuellar received a Ph.D. from Centro de Investigaciones en Óptica, Mexico, in 2009. Currently, he is working as researcher at Centro Universitario de los Lagos, University of Guadalajara. He has been a visiting researcher at Instituto Potosino de Investigación Científica y Tecnológica (IPICYT), Mexico; Faculty of Radiophysics of Lobachevsky State University of Nizhny Novgorod, Russia; and St. Mary's University, San Antonio, TX, USA. He has edited two books and published six book chapters and forty-two high-impact publications. He served as an academic editor for the journals *Complexity* and *Frontiers in Applied Mathematics*. He has also participated as a reviewer for scientific journals. His research interests include the study, characterization, and dynamical behavior of fiber lasers and electronics circuits.

Contents

Preface

Optical fibers are of great interest for scientific and technological applications, especially in communications systems.

In the first section on technology, Chapter 1 discusses the technology that has been developed to generate better optical fiber products with increased transmission speed, such as multi-core fibers. Increased transmission rates can be achieved via the use of space division multiplexing, allowing for an increase in the information in one optical fiber. Of course, a multi-core fiber must be coupled with specific electronic and optoelectronic devices to increase its performance.

A wide variety of fiber devices can be created by adding special coatings on tapered sections of optical fibers. Chapter 2 presents the fundamentals for the fabrication of tapered optical fibers coated with functional polymers. Here, the required aspects of light propagation in tapered sections of optical fibers are introduced and the relevant parameters enabling light interaction with external media are discussed. A special case of interest is the addition of polymeric coatings with prescribed thicknesses in the tapered sections allowing for adjusting the light propagation features; using liquid polymer coatings with varying thicknesses along with the taper profile that can be tailored for tuning the transmission features of the devices. The chapter also presents a methodology for obtaining coatings with predefined geometries whose optical properties will depend on the polymer functionality.

Optical fiber networks are the standard applied for high-bandwidth customers. Various access technologies to business networks with very high bandwidth to access networks for buildings and individual consumers have emerged. Chapter 3 discusses the optical network connections inside buildings for commercial and domestic users, including the use of optical glass fibers or/and polymeric optical fibers in different network topologies in connection to high-speed actual WIFI technologies.

The second part of this book is dedicated to the application of optical fibers in communications. Chapter 4 deals with coded modulation and impairment compensation techniques in optical fiber communication. Probabilistic shaping is a new coded modulation technology that can reduce transmission power by precoding, reducing the bit error rate, and improving communication rate. In this chapter, a probabilistic shaping 16QAM modulation scheme based on trellis-coded modulation is proposed. Experimental results show that this scheme can achieve better optical SNR gain and BER performance. On the other hand, to meet the demand of transmission rate of next-generation high-speed optical communication systems, multidimensional modulation and coherent detection are sufficiently applied.

Nowadays, due to the renewed demand on data bandwidth imposed by the upcoming capacity crunch, the optical communication (research and industry) community has oriented its effort to space division multiplexing (SDM), particularly mode division multiplexing (MDM). This is based on separate/independent and orthogonal spatial modes of optical fiber as data carriers along with the optical fiber. Chapter 5 reviews the potentials of harnessing SDM as a promising solution for next-generation global communications. The study is focused on different SDM approaches and specifically addresses MDM (different modes in optical fiber).

Because of their versatility in different areas of science and technology, optical fibers have many important applications, as outlined in the last part of the book. Chapter 6 discusses the use of optical fiber tweezers for the assembly of living photonic probes, which are versatile tools for optical trapping and manipulation that have attracted much attention in cell trapping, manipulation, and detection. The assembly of living cells using optical fiber tweezers has attracted significant attention. Advanced achievements have been made on the assembly of fully biocompatible photonic probes with biological cells, enabling optical detection in a biological environment in a highly compatible manner. Living photonic probes can be assembled by trapping and assembling multiple cells using optical fiber tweezers. These photonic probes exhibit high biocompatibility and show great promise for bio-applications in bio-microenvironments.

Optical fiber sensing research has been extended to the area of detection of microorganisms such as bacteria, viruses, fungi, and protozoa. The validation of optical fibers in bio-sensing applications can be observed from the growing number of publications. Chapter 7 provides a brief picture of optical fiber biosensors, their geometries, and the procedures for their development.

Integration of optical fibers with optoelectronic devices is a very important application. In terms of communication systems, they are integrated with one or two technologies, as is the case for automation, image processing, and embedded systems. Chapter 8 is dedicated to recent results of artificial intelligence (AI) and photonics integration. Nowadays, photonics are used with AI to facilitate ultra-fast AI networks to offer a novel class of Information Processing Machines (IPMs). The chapter demonstrates the implementation of photonics for AI utility and AI for photonics. In this category, a dual-core photonic crystal fiber (PCF) integrated with AI is proposed, which serves to identify infected human cells. This proposed design of PCF is providing relative sensitivity and confinement loss in an optimized manner with the impact of AI.

The numerical approach for characterizing some applications is a very important tool. The final chapter of this book considers differential equations and analyzes methods with Opto-Electronic Oscillator (OEO) modulation with direct and external modulation. The complexity of both approaches is related to the non-standard way of description of the nonlinear method modulation for the internal (direct) structure and the utilization of the specific Mach-Zehnder modulator for the first stage of external modulation. The purpose of the presentation is to consider the main features of OEO as a low-noise generator. This includes consideration based on the study of differential equations, the study of transients in OEO, and the calculation of phase noise. It is shown that different types of fibers with low losses at small bending radii can be used as a FOLD in OEO.

This book was made possible thanks to the effort and dedication of researchers worldwide that want to share their knowledge with the world, facilitating the spread of different possible innovative experiments and technological developments. In that sense, this book covers a good diversity of topics from fiber optics technology development to communications systems and some applications for scientific research and technological devices.

Guillermo Huerta Cuellar
Department of Exact Sciences and Technology,
Universitary Center of Lagos,
University of Guadalajara,
Lagos de Moreno, Mexico

Section 1

Optical Fiber Technology

Chapter 1

Multi-core Fiber Technology

Muhammad Irfan Anis and Hamdan Ali

Abstract

Traditional single-mode fiber capacity issues will be mitigated by using space-division multiplexing in future 5G, IoT, and M2M networks. Multi-core fibers are expected as a good candidate for overcoming the capacity limit of a current optical communication system. This chapter describes the recent progress on the Multi-core fibers technology for the application of high capacity space-division multiplexing to be utilized for long-distance transmission systems. Further various optical approaches that enable key functions are discussed, including SDM MUX/DeMUX, switches, transceivers to enable next generation optical network. Moreover, issues like crosstalk, non-linearity is a potential limitation on the achievable data-rates in optical fiber transmission systems using multi-core fibers will be discussed.

Keywords: All-optical signal processing, Crosstalk, Fiber capacity, Optical Network, Space division multiplexing

1. Introduction

Internet traffic infrastructure is underpinned by optical transmission systems and networks [1]. However, continuous supply of new services like video on demand, Virtual Reality and Augmented Reality have rising data volumes needed to satisfy the requirements of industry, academics, governments, and people presents new challenges to optical communication infrastructure [2]. Fiber-optic communication systems based on conventional single mode single core fibers (SMF) are almost saturated due to amplifier bandwidth, nonlinear noise, and fiber fuse phenomena due to this it's capacity consumption will be beyond capacity limitations by the year 2022 [3]. As a result, researcher has led to a steady push for new, higher bandwidth optical fibers that can replace the SMF.

Space division multiplexing (SDM) methods is one of the potential capacity expansion strategies for an optical transport network. It is transmission fibers that enable concurrent parallel data transmissions on multiple cores in a single cladding or several cores inside a single core to improve speed and data rate [4]. The first demonstration of an SDM link consisting of standard cladding diameter surpassing the typical size of 125 μm 7-core MCFs, highly efficient MC-EDFAs, and MCF connectors transmission above 100 Tbit/s across a 316 km has occurred and considerably greater capacity tests such as over 1 Pbit/s and 1 Ebit/s·km were performed using single mode multi-core fiber [5]. SDM fibers are defined as multi-core fibers (MCFs) and few-mode fibers (FMFs) or multi-mode fiber (MMF) [6]. MCF and multi-mode fiber technologies provide for additional fiber capacity proportional to the number of cores and modes per fiber [7]. Single-mode cores, contained in a shared cladding, are employed independently in the former. An FMF has one core, which enables several optical modes, each of which may transmit data independently. MCF is a promising technology for providing enormous bandwidth and capacity with regard to information [8]. The MCF help facilitate the data transmission and the transmission of power in high power devices. Multi-core fibers have many positive attributes over conventional fibers: they have significantly decreased core separation and are very regular when compared to free-standing fibers, and they also provide a monolithic package with several fiber features [9]. Moreover, multi-core fibers that allow a few-mode core to combine the fibers results in an extra 100 optical channels in each transmission, as well as throughputs of over 1 Pb/s [10]. Recently MCF-based fiber-optic transmission, a capacity of 1 Pb/s per 32-core fiber has been achieved [11].

Multi-Mode-Multi-Core Fiber (MM-MCF) significantly increases the number of spatial channels to 114 or more, and transmission of 10 Pbit/s was achieved utilizing this multi-mode MCF. Despite these benefits, the MCF may have limitations such as crosstalk (XT), non-linearities, dispersion, and so forth. Over long distances, the accumulation of MCF crosstalk may be the most limiting issue influencing the performance of an optical communications system. As a consequence, in recent years, research in this field has been driven by the development of ultralow crosstalk MCFs [12]. The impact of XT on MCF system capacity and range has recently been studied [13]. However, the results vary with modulation format and transmission reach, leading to the general notion that different network applications, from short-range to ultra-long haul, need different MCF designs. Standardization and mass production are essential for widespread commercial usage of emerging technologies like MCF. An XT standard per unit length of 55 dB/km has been proposed [14]. In an MCF system, the performance penalty must be evaluated against a non-XT system, regardless of unit size (i.e., a fiber bundle instead of an MCF). Capacity and reach penalties are required. Calculation on an optimal MCF core density for long-distances-independent crosstalk specifications have been done. The crosstalk process was originally described in [15], although the majority of crosstalk on a fiber is continuous, it is at discrete places where crosstalk amplifies the most, when core matching circumstances occur. Since the locations and phases of these sites may change randomly, crosstalk in MCFs follows a random chi-square distribution with 4 degrees in time and wavelength [16].

Using MCF's nonlinear distortions for power-over-fiber operations poses a number of challenges. The structure of MCF, which enables for high-power signal transmission via the fibers, has lately received attention. It is recommended that image processing be used; therefore, the present limitation of single mode fiber must be overcome [17].

A 7-core MCF with reduced inter-core crosstalk was used for trans-oceanic transmission. Using MCF and a spectrum efficient modulation scheme, 201 x 100 Gbit/s transmission across 7326 km produced a capacity-distance product surpassing 1 Exabit/skm [18]. These systems propose the MCF as one of many optical transmission techniques.

To transmit 52.2 Tbit/s across 10230 km, the CDP for SM-SCF transmission is 534 Pbit/s/km. The transmission rate was 1.03 Ebit/s km/h [19]. A preliminary test using seven spatial channels and PDM-QPSK yielded 53.3 Tb/s [20]. Using 8 spatial channels with PDM-8PSK, the capacity was 83.33 Tb/s. MMF allows transmission distances up to 1200 km (3 spatial modes x 40 Gbit/s DP-QPSK) and 13.9 Pbit/skm (CDP with MMF)[21]. Uncoupled MCF allows for considerably longer transmission. The 1500 km transmission used a propagation-direction interleaved design to minimize interference between neighboring cores [19].

This chapter investigates MCF-based novel technologies for creating next-generation optical networks. We first examine the roadmaps towards optical fiber and examine why there is a need for MCF. We also highlight the newest reports' in MCF paradigm covering design and application. Looking into the main technology as a key functional building blocks for next generation optical communication. Next, we demonstrate the experimental setup of MCF. In last section describe the MCF limitation before conclusion.

2. The roadmaps towards fiber optics

The global proliferation of hyper photonics, intelligent photonics and frontier wireless communication need an increasing data capacity of tens of percentage each year. Services including IoT, M2M, sensor networks, and linked vehicles will need even greater bandwidth to expand capacity and find more efficient connections via high-speed optical fiber networks, as **Figure 1** shows the innovation technologies. As a result, the forthcoming growth in data transmission will exceed the SMF's maximal transmission capacity due of its low loss and optical amplification of the transmission window. Traditional SMF cannot be ignored in DWDM transmission systems with Raman amplification [23]. An increase in optical infrastructure is required to meet capacity constraints. Introducing extra optical fibers and cables is considerably simpler when contemplating an alternate technology. The future capacity constraint may be averted simply by creating more SMFs. Thus, construction/renewal of the physical infrastructure would be required, which would add to the total expenses. SDM, a fifth physical dimension, may supplement time, wavelength, frequency, and polarization multiplexing, thereby easing future capacity problems [24].

Figure 2 shows the progress of cable density. 400-pair copper, 400-fiber ribbon, and 400 rollable fiber ribbon cables are indicated by black, blue, and red dots. A solid green line indicates the numerical limit for a hexagonally packed 250 mm fiber bundle with a 2 mm cable sheath [24].

The two spatial dimensions of mode and core are used in the design of optical fiber. MCF stands for multi-core fiber division multiplexing, while MMF stands for multi-mode fiber division multiplexing. For understanding description of a 2D representation of modes and cores inside optical fiber shown in **Figure 3**. With proper use of modes or cores, it is possible to surpass the present geometric limit of conventional optical fiber cable. Using the modes and cores in tandem will almost triple the spatial multiplicity. This recent study concludes that 6-mode and 19-core fiber can provide over 100 spatial channels [25]. It's required that a complicated transmission strategy be used because of the mode coupling and/or mode-dependent transmission properties in optical fiber. Also, MCF has been constantly studied and was pioneering in [24]. MCF is especially capable of using the newest

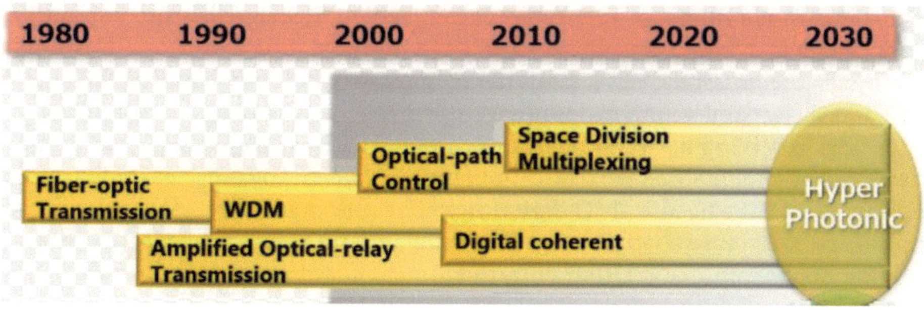

Figure 1.
History of optical network innovation technologies [22].

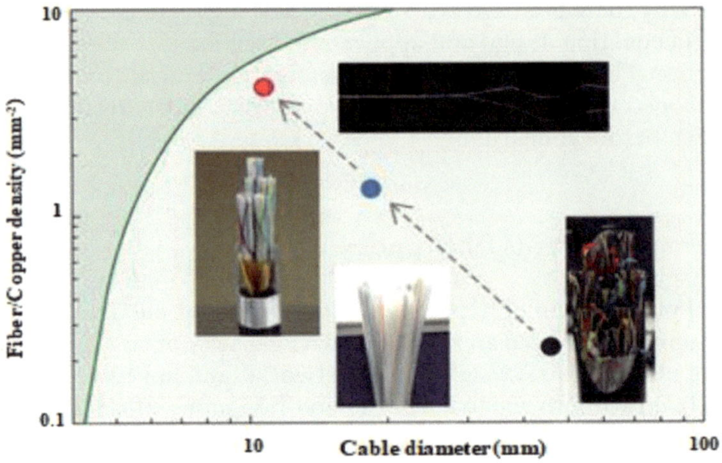

Figure 2.
Evolution of communication cable density over time [7].

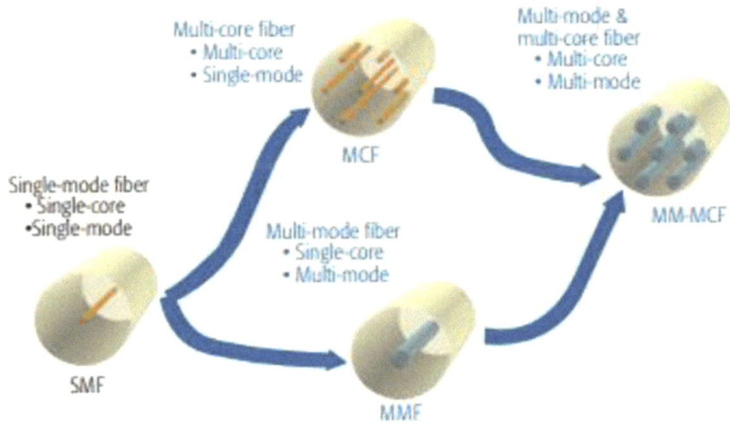

Figure 3.
Schematic image of the two spatial dimensions of mode and core in optical fiber [24].

single-mode technologies. Here, we'll talk about MCF and the capacity of MCF as an SDM transmission medium.

Signal mixing and propagation time skew are two important characteristics of SDM fibers that have a direct effect on transmission performance. SDM fiber groups are provided by two mixing levels. First, there are uncoupled MCFs (UC-MCFs), few-mode fibers, or few-mode MCFs, which mix signals from several spatial channels during transmission and have minimal inter-core coupling to minimize inter-core crosstalk (XT). Both the first group randomly coupled MCF (RC-MCF) and the second group randomly coupled MCF (RC-MCF) have significant random mode mixing between modes (RC-MCF). UC-MCFs have greater spatial channel density than SMFs with traditional transceivers. To compensate for random coupling, RC-MCFs need MIMO digital signal processing (DSP). However, random coupling may decrease spatial mode dispersion and therefore the MIMO DSP's computational complexity. Other core/mode-dependent restrictions are also prevented by random coupling [26, 27].

MCF with a 125 mm cladding diameter is needed to start MCF tech. A 125 mm cladding diameter MCF design that is optically compatible with current SMF.

	No of cores	Core layout	Cladding diameter [μm]	Cut off wave length [μm]	Mode field diameter [μm] @nm	Affective area [μm²]@nm	Crosstalk [/km] @nm	Wavelength band	Applications	Parameter/feature	Ref.
Multi-core Uncoupled-type	7	⬡	150	≤1.51	9.8@1550	80@1550	$6.0x10^{-9}$ @1625	C ~ L	LH	Ultra-low Crosstalk	[29]
	7	⬡	188	≤1.47	12.2@1550	124@1550	$8.0x10^{-7}$ @1625	C ~ L	LH	High Optical SNR	[30]
	31	⬡	225	~1.47	n/a	57@1550	9.3x10	C ~ L	LH	High SSE	[31] [32]
	4	□	125	≤1.19	8.6@1310	—	$5.0x10^{-5}$ @1625	O ~ L	LH	Standard diam. Cladding+ Full wavelength band	[32] [5]
	8	○	125	≤1.24	8.4@1310	—	$3.2x10^{-7}$ @1310	O	SR	Standard diam. Cladding +8 cores	[33]
	8	2x4	180	≤1.20	8.4@1310	—	≤$6.3x10^{-5}$@1550	O ~ C	SR	Si photonics TRx mounted	[34]
	4	1x4	98x200	≤1.34	9.7@1550	n/a	$3.0x10^{-4}$@01550	C ~ L	SR	Non-circular clad.	[35]

	No of cores	Core layout	Cladding diam. [μm]	Cut off wave length [μm]	Mode field diameter [μm] @nm	Affective area [μm²] @nm	Spatial mode dispersion [ps/√km]	Wavelength band	Applications	Parameter/feature	Ref.
Multi-core coupled-type	3	△	125	~1.35	—	129@1550	30	S ~ L	LH	4,200-km MIMO transmission achieved.	[35]
	4	□	125	≤1.47	—	112@1550	3.1	C ~ L	LH	Has set low SMD and low-loss records for SDM fibers. 10,000 km transmission achieved.	[36]

Table 1.
Representative examples of reported MCFs [28].

Table 1 depicts conventional MCFs provided by Sumitomo Electric, as well as prototype MCFs created by the company via joint research.

2.1 Need for multi-core technology

The restricted bandwidth of low-loss transmission and optical amplification, as well as transmission power limitations due to fiber non-linearity, make expanding a single optical fiber's transmission capacity challenging. A growing need for higher-capacity optical fiber communications **Figure 4** shows high-capacity optical fiber transmission test results. Due to the limitations of single-mode single-core fiber, the maximum capacity is 100 Tbit/s. This high-capacity fiber capacity was achieved using MCF. SDM plus MCF or MMF may be able to outperform single-core fiber transmission systems [12, 26, 29].

3. MCF paradigm

This section investigates the design and the achievements in MCF technology, which seem promising in the short term yet have certain unknown risks.

3.1 Design of MCF

Due to fiber bandwidth depletion and the development of SDMF as a possible alternative to address extra capacity, the spectrum capacity of SSMF is nearing its end. The long-term goal of SDM is to increase the number of fiber cores, guided modes, or both. MCF has many cores in a single optical cable. The core of a conventional single-core fiber is positioned in the center of a 125-m diameter cladding, limiting design freedom. The MCF's success is dependent on more than simply the number of cores. MCFs enable the designer to optimize core design, the number of cores, core arrangement, outer cladding thickness, and cladding diameter in terms of optical and mechanical properties. Fiber design is required based on the application because desirable features differ. SMFs currently have a single fiber core surrounded by 125 mm cladding and coated. Greater cores with the same cladding or larger core diameters allow for more fiber capacity [12]. Adding cores to the

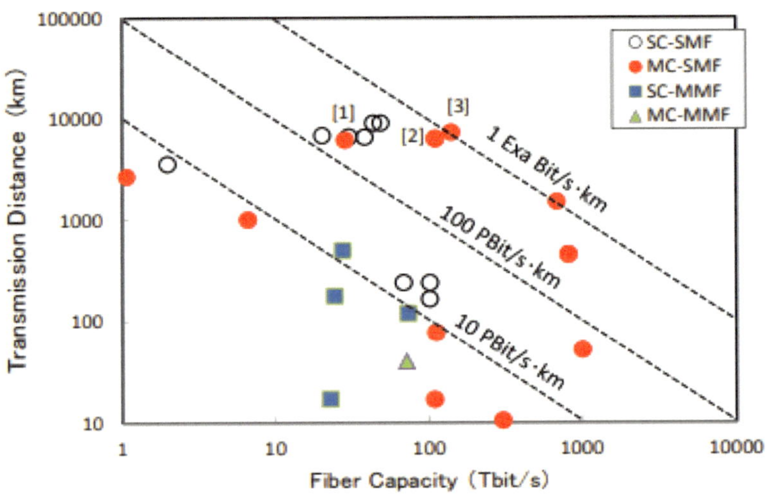

Figure 4.
Recent reports on high capacity transmission [19].

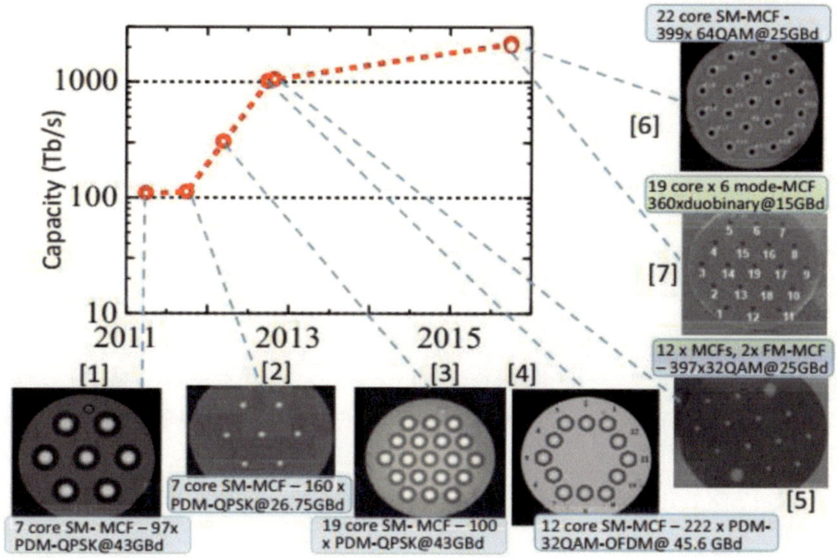

Figure 5.
High-capacity transmission experiments using SM and FM-MCF [12].

cladding may improve capacity, but it may need changes to the transmission system architecture.

SM–MCF transmission experiments on FMFs have shown fibers with as many as 32 and as little as 45 modes. Adding multi-core fibers that enable a few-mode core to join the fibers results in an additional 100 optical channels in each transmission and throughputs of more than 10 Pb/s [30]. The high-capacity experimental transmission utilizing SM and FM-MCF is shown in **Figure 5**.

A random coupled MCF is one that is MCF if XT is compensated by MIMO DSP. Even though the core is simple, the paired MCF is denser. Furthermore, random coupling in the connected MCF prevents the emergence of nonlinearity impairment, SMD, and mode-dependent loss/gain. In long-distance point-to-point communication, coupled MCF is utilized. Nokia Bell Labs, Sumitomo Electric, and Sumitomo Corporation collaborated to develop and launch new fibers [31].

Few-mode (FM) MCF fiber is a kind of uncoupled MCF (MCF with fiber coupled together) designed for mode-multiplexed transmission. KDDI Research, Inc. received a prototype 36-core fiber created in cooperation with NICT and Yokohama National University. The most recent accomplishments include a 19-core fiber that can be used in the whole C + L bands (1530–1625 nm) for long-distance communications [37]. This fiber achieved 10 Pbit/s per optical fiber in an experiment performed by KDDI Research [30].

The MCF has the potential to enhance data and power transmission for high-power devices. PoF, on the other hand, need MCF due to its nonlinear aberrations. Inside MCF, an eye-catching power transmission capacity was recently discovered. The placement of the cores has an impact on the MCF's performance.

Multi-core fiber architectures such as triangle, ring, square, rectangle, and hexagon were developed after analyzing the number of cores, pitch, and power spectrum. Many people are interested in the MCF fiber-optic structure and the question is how it allows the transmission of powerful signals. One-mode fiber has a lower limit imposed by the MCF and is currently limited by MCF analysis and picture processing. The placement of the cores influences the performance of the MCF.

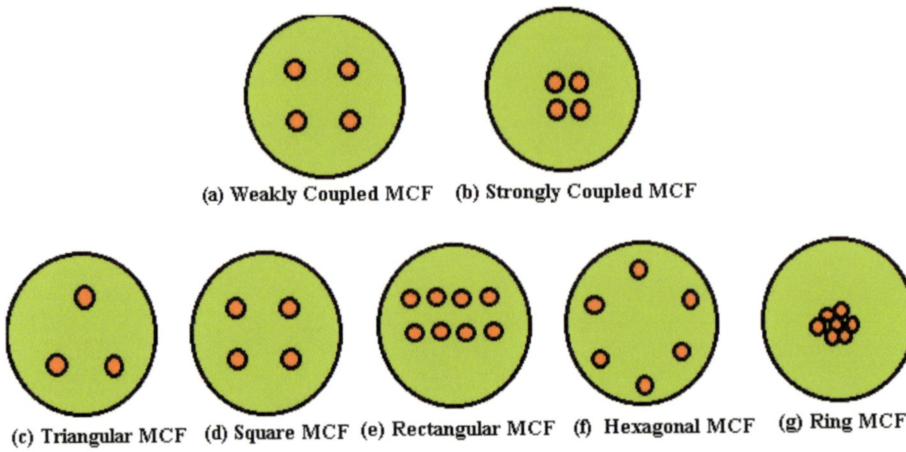

Figure 6.
Structure of multicore fiber with coupling region and with different pattern [17].

For MCF, strong and weak coupling are illustrated in **Figure 6**. Strongly linked MCF has the smallest core-to-core distance, whereas weakly coupled MCF has the most core-to-core distance. The size of the core may vary with the pitch of the core. The effective area (A_{eff}) relies on the number of cores (as shown in **Figure 6**), and their configuration influence the output receiving power. MCF contains several cores with enlarged effective areas, resulting in minimal dispersion and bending losses. The suggested solution to fiber bending losses included four air core MCF. **Figure 6** shows the five distinct MCF structures.

To provide long-distance reliable signal transmission, the XT must be less than -30 dB/100 Km [32]. To get ultra-low XT levels, make changes to MCF structures, such as trenching around the cores. Essentially, trenches are refractive index profiles with lower refractive index than the core and cladding. Trench-assisted method is one of the noteworthy techniques that lowers the coupling between the adjacent cores, therefore helping to minimize existing crosstalk.

With an MCF, if the number of cores in a restricted cladding area grows, crosstalk suppression becomes a problem. XT in MCFs is decreased by decreasing the coupling coefficient between cores. The underlying design, with strong containment of modes is critical to suppressing the mode coupling coefficient. For a higher A_{eff} and lower nonlinear noise, you may choose for a higher-index core with a smaller diameter.

It has three important geometrical features, as shown in **Figure 7**. The outer cladding thickness (OCT) is the distance between the outer core's center and the cladding's perimeter. Optical fiber mechanical reliability is strongly linked to cladding diameter D. A higher D value increases MCF deformations before collapse. Inter-core XT may be reduced by adjusting core and rod radius, cladding and rod relative refractive index differences, and core-to-core distance.

3.2 Application

The MCF technologies have been gradually increasing, and now we can see feasible commercial uses for the technologies. Practical use of MCFs will likely occur in near future due to continuous MCF development [28]. **Figure 8** illustrates the whole growth stages of MCF technology. Larger applications in the network such as metro and core may provide challenges to MCF implementation, because they need a complete suite of network components other than the MCF and cable

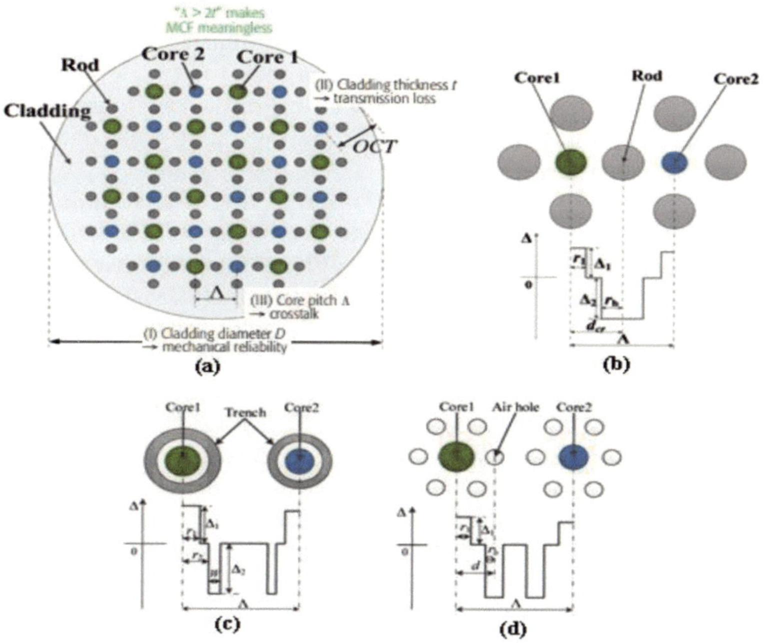

Figure 7.
(a) A 32 core schematic structure shows three key geometrical parameters in MCF. Schematic diagrams of (b) core index profile of heterogeneous rod rod-assisted 32-core fiber (c) trench-assisted (TA) profile, (d) hole-assisted (HA) profile [33].

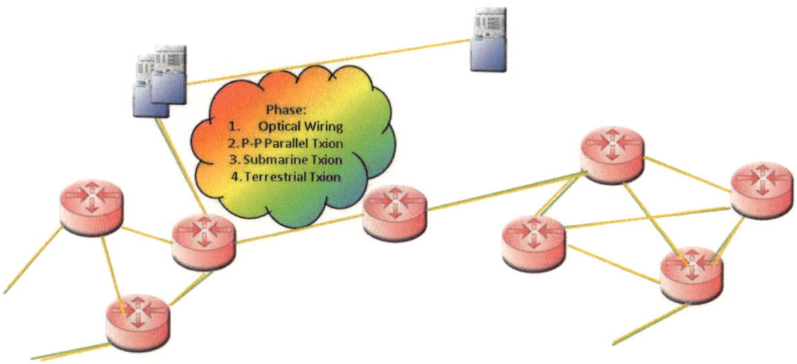

Figure 8.
Schematic image of expansion phases of MCF technology [24].

(e.g., optical amplifier, and node management technologies). To a large extent, central offices and/or data centres are the primary target for MCF technology since they are maintained and/or optimized separately by experienced operators, making it simpler to upgrade current network components. Here, compatibility with traditional SMF is very essential, especially for connection. Then, in the second deployment phase, we use MCF P2P and/or parallel transmission technology. MCF requires a flexible connection to the optical subsystems. Submarine transmission systems may offer more promise because of their use of the newest technology, and SDM might possibly achieve a power-efficient transmission system [34]. We have finally achieved flexible and dependable SDM nodes [24].

A FORECAST predicts that data center network traffic would increase at a 25% CAGR, with most traffic (about 70%) staying within the data center [35]. Modern data centers utilize dense non-blocking topologies with point-to-point optical transponder connections. Either on end servers or in slots of electronic switching devices, data is electrically switched. This resource-intensive paradigm may cause future data center scaling problems. Modern data center networks are increasingly relying on optical technologies like hybrid electrical-optical switching (HEOS). Recent demonstrations of a time-slotted pSSC and core-joint SDM optical switching system for edge applications [36].

SDM may improve network capacity by multiplexing SMF strands, multicore fiber cores, or even each mode of a few-mode fiber using MIMO digital signal processing [38]. Spatial Division Multiplexing-Elastic Optical Networks will then be the future of optical transport and data center networks (SDM-EON) [39]. MCF front haul multiplexes MIMO signals onto a single cable, enabling multiple optical data streams to be transmitted simultaneously. The MCF may also provide a single optical data signal to each antenna element, with varying delays and phases. Multi-antenna systems need MIMO and beamforming. The MCF front haul uses MIMO signals to transmit multiple optical data streams at the same wavelength. The MCF also uses optical data transmissions with variable phase or time delays to each antenna element. 5G systems need MIMO and beamforming capabilities. **Figure 9** shows a multi-antenna MCF-based RRH-to-remote-site connection with optical beamforming and/or digital MIMO capabilities. These methods enhance system performance.

The system capacity and accessible user bitrate may be enhanced by multiplexing MIMO data streams. A 22 MIMO LTE-A transmission using MCF technology was evaluated early. This research adds 44 MIMO transmission supported by a 4-core fiber capable of feeding four AEs concurrently. The M(22) arrays allow 5G systems to control multiple groups of four AEs. MCF may be used to reduce the size and complexity of beamforming systems. A same data signal is supplied by four separate AEs with varying delays, as shown in **Figure 9**. MCF aligns all optical lines in the beamforming system, simplifying the network [40].

In [36], Making fiber bundles revealed a 19-core MCF. A 7-core MCF Micro-lens array (MLA) claims 47.8 dB return loss and 0.87 dB insertion loss. Tapers were made, then cut apart to make fused fiber. Non-mode-selective, in which the modes spin on the device itself, and mode-selective, with minimal unitary rotation between modes. The most common components were lamps, phase plates, PLCs, and then mode selective PLCs. Ultrafast laser inscription can produce low loss 3D waveguides in conventional optical glass for MCFs. 3 mode FM-MCF fiber with average IL 0.92 and homogeneity 0.1.

SDM enthusiast offered considerable flexibility in fiber light mixing, integrated sensors and controllers. MCF technology has also been used to construct optical fiber sensors, which make them excellent for industrial applications. High

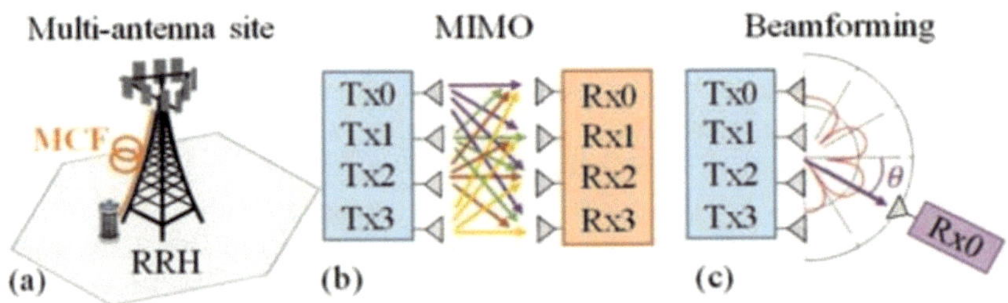

Figure 9.
(a) A multi-antenna site application scenario in which the RRH is linked to the antenna components through MCF. Examples of four-antenna systems with (b) spatial multiplexing using 44 MIMO (4 distinct data streams coded in 4 layers) and (c) 41 beam forming (4 antennas transmitting the same data with different delays) [40].

temperature sensing to 1000°C using MCF optical sensors with a typical temperature sensitivity of 170 pm/°C Mach–Zehnder (MZ) interferometers may be fabricated by employing MCFs since the slopes of the resultant interference peaks are steeper. Until quite recently, many MCF optical sensors used inefficient methods to throw light into the multi-arm MZ, causing substantial losses for QI processing [41] use novel tapering methods to construct the multi-arm MZ directly into a specially built MCF.

4. Key functional building blocks for next generation optical communication

Aspects of future network and technology MCF also requires FI/FO devices, connections, amplifiers, and integration technologies. This section covers the optical network equipment relevant to this research.

4.1 Space division multiplexer/demultiplexer

MCF technology uses SDM MUX/DEMUX. There are now numerous options, each with its own footprint, cost, capabilities, multi-mode affinity, etc. An SDM multiplexer or demultiplexer effectively links light between SMF fibers and SDM fiber modes or cores. Spatial MUXs are needed for SDM studies and may be used to link SMF and SDM networks in the future. This connection has been suggested in many creative ways. Direct and indirect coupling methods are widely classified. The optical signal is fully confined inside a waveguide during connection. **Figure 10** shows two typical layouts. As illustrated in **Figure 10**, the SMF cladding diameter is tapered to splice a bundle of SMFs to the SDM fiber. A photonic lantern is made by compressing MCF or SMF cores into an FMF. Alternatively, an unneeded

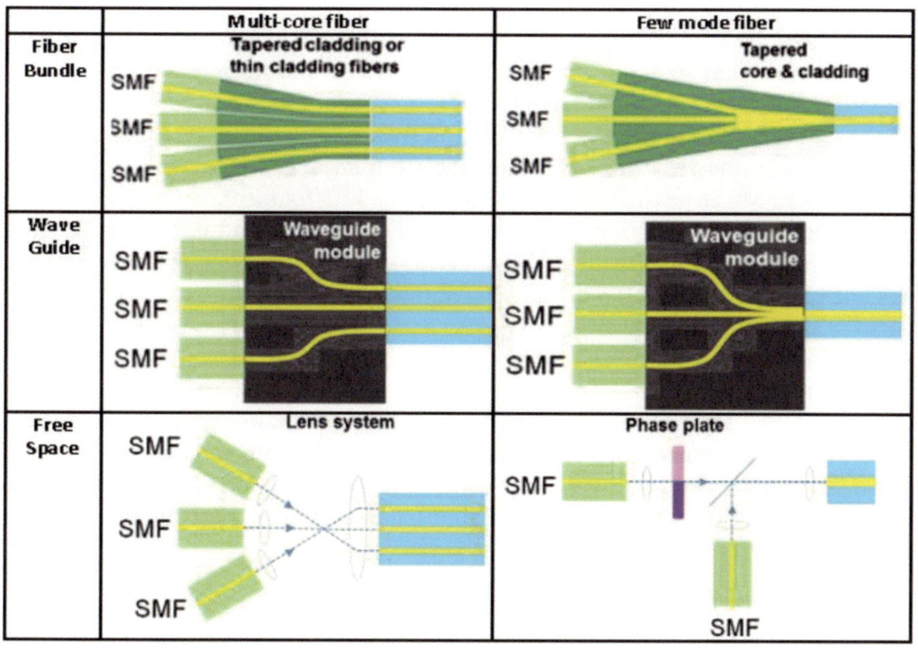

Figure 10.
Fiber bundle, waveguide and free space MUX for MCF and FMF respectively [12].

waveguide may be used. Inscribed light-guiding cores on a tiny glass block form the waveguide [12].

Figure 10 shows the inscribed cores in the waveguide output plane separated by the MCF core separation. Input and output SMF arrays are connected using UV-cured glue. In addition to fiber producers, Chiral Photonics uses the fiber bundle technique often. Optoscribe's commercialized 3D waveguide technology is easy to incorporate with photonic integrated circuits (PICs). Indirect coupling uses bulk optics like lenses and prisms [12].

4.2 Transceiver

Transceivers are devices that combine the operations of a transmitter and a receiver into a single unit. They connect the network to a computer module in both directions. Their duty is to generate optical signals and then convert them to electronic data. Aside from the form factor and connectors, the optical and electrical properties are significant factors to consider throughout the selection process. The transmitter determines the wave properties of the transmission. The wavelength, spectral breadth, and transmission power are all critical parameters. Other transmitter characteristics include wave modes, reflections, and so on. To receive an optical signal, the receiver must be set to the proper wavelength. Furthermore, the signal's polarization and power must be compatible. Photo-diodes and other light-sensitive semiconductors transform optical impulses into electrical signals that may be monitored for data extraction. The received power must be within the detector's permitted range. If the power is too low, it is impossible to differentiate between signal and thermal noise, resulting in a poor signal-to-noise ratio. The detector gets overwhelmed if the received power is too high. Modulations of the luminous flux are not detectable. Overloading may permanently harm the detector and should therefore be avoided [42].

4.3 Connectors

Connectors are devices that connect optical wires. Connectors are required for SDM systems such as fusion splicing in terrestrial and submarine trunk networks. For a variety of cable types and transmission methods, many connection types have been created. Due to the fiber break, the link transmission is lossy. Lenses, end polishing, and forms are utilized to decrease attenuation. M-type connections were used for 7-core MCFs with an IL of 0.13 dB and a 500-fold improvement in MTBF. A multi-fiber MPO connection with over 40 dB return loss and 0.85 dB IL. A 7-core MCF connection with a return loss of 45 dB and an MPO connector for four 7-core fibers with a return loss of 0.3 dB are also shown in the study.

4.4 Amplifiers

Erbium-doped MCF amplifiers may be constructed utilizing separate pump lasers. Sharing pumps across multiple cores enhances power efficiency. Another approach is to pump the MCF's cladding, which is outfitted with multi-mode lasers. To achieve greater efficiency than an array of SSMF EDFAs, more power must be injected [43].

4.5 Switches

A network's heart is comprised of switches. Switches manage signal paths between nodes. In traditional copper networks, this routing is based on data packet IDs. Routing in optical networks, on the other hand, may be based on physical

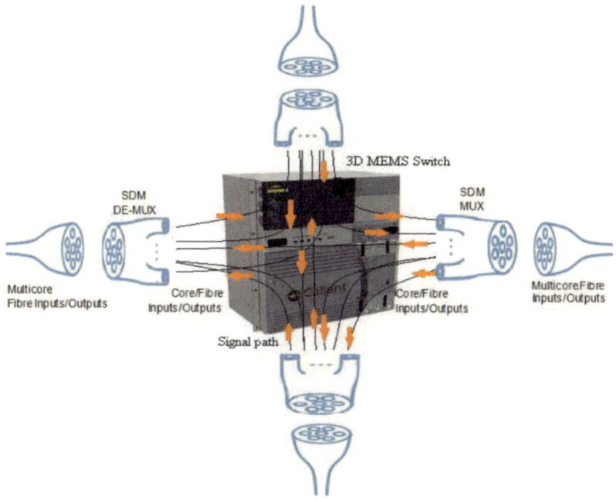

Figure 11.
Core switching with spectrum contiguity within an optical network that uses multi-core fiber transmission.

signal properties. This may be the wavelength in WDM or the core in multi-core fiber transmission, for example.

Figure 11 is an example of a signal route in a switch. All the linked wires has seven cores. Every signal has its own core. When routing a signal, it may be done dynamically or statically. The physical routing technique is switch-dependent, which means that different switch types will have different methodologies. Due to physical coupling effects between the various cores, the range of multi-core fiber technology is limited to km. The limited flexibility of single-core fibers is distinct from that of multi-core fibers [44].

Low-speed SDM optical switches are already promising technologies, and preliminary work on SDM optical switches based on MEMS or LCoS technologies has been done. This will benefit WAN networks that need high-layer packets to be routed directly into the optical domain. Optical fast-switching networks have never achieved broad adoption owing to building difficulties, quicker signal degradation, and lower-cost electronics.

These switching granularities arise from the spatial component of SDM networks: Space granularity (joint switching) is needed when all modes intermix. Fibers like FMFs have full wavelength granularity in fractional space. Recently, several papers on SDM optical switches have appeared. In [36], a heterogeneous WSS switches spatial channels in an FMF, SSMF array, and SC-MCF. [36] claims a three-port four-core MCF WSS for SDM with 34 dB crosstalk and IL under 2.2 dB. Reference depicts a silicon PIC with a 7 7 switching matrix (MZIs) with an insertion loss of [4.5, 7.0] dB. Acoustic-optical crystals may also be used to create SDM optical switches. **Figure 12** depicts a CJ-AOM switch for 7 spatial channels. A 10 second switching time with an insertion loss of 10 dB.

There are spectrum resources in each core in SDM-EONs. All Spectrum slots are created equal. Following the spectrum contiguity requirement implies the whole service must utilize the same spectrum slots along the lightpath. To keep spectrum continuity constraint in a fiber, service spectrum slots must be continuous in the spectral dimension. The OFDM method should be used for each core to enhance spectrum efficiency. Spatially and spectrally resolved optical switching fibers are made as shown in **Figure 13**. In the optical switching fabric, core, fiber, and spectrum switching may be accomplished, which enables flexible channel addition, removal, and wavelength-level granularity channel switching. A transceiver pool supplies the necessary sub-transceivers for the different communication

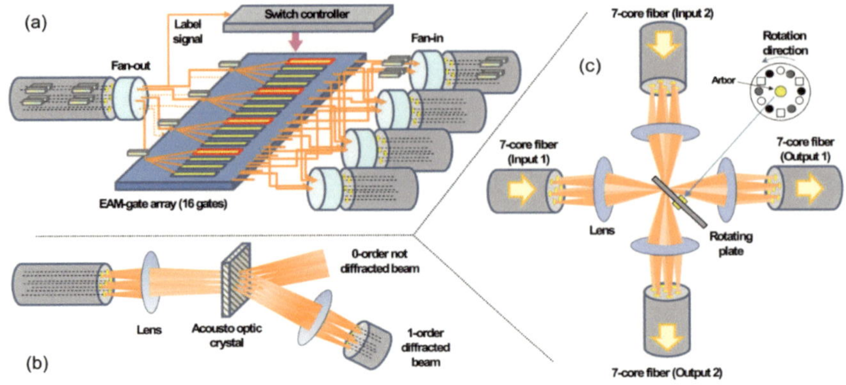

Figure 12.
(a) Core-joint electro-absorption switch diagram, (b) core-joint acousto-optical modulator switch diagram, and (c) core-joint mirror switch diagram [36].

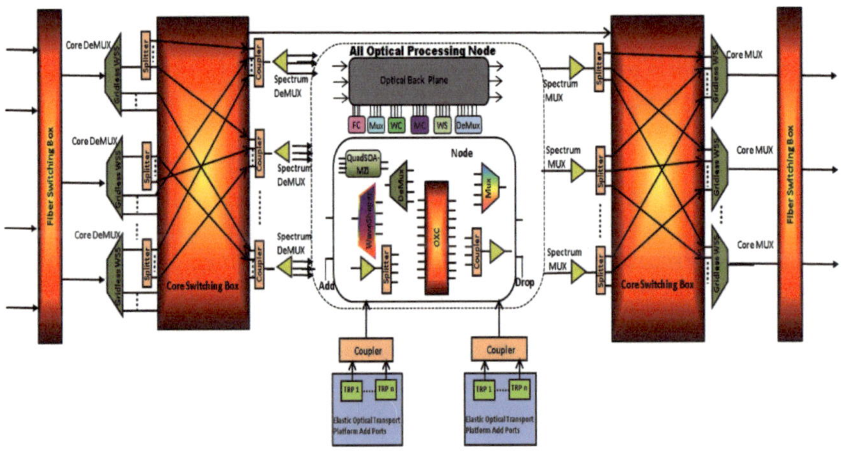

Figure 13.
Spatially and spectrally resolved optical switching fabric.

requirements. To overcome the spectrum contiguity restriction, spectrum slots in the switch fabric may be swapped between various cores. To summarize, it is possible to flexibly move signal cores without losing spectrum.

5. Limitation

Researchers all around the globe are working to reduce problems in order to attain ultra-low signal distortions in fiber optic technology [45]. However, high-capacity transmission systems place extra importance on network reliability [46]. It has gotten a lot of attention as a promising technique for dealing with the capacity limitations that are associated with single-core SMFs and cable size limitations like in datacenter networks and Passive Optical Networks (PONs) (which require high fiber count and high-density). Also, MCFs provide redundant signal lines and primary signal lines, which enable them to construct extremely dependable networks [47].

5.1 Cross talk

MCF is currently actively researched for SDM. SDM-based long-haul transmission requires low-crosstalk (XT) architecture. MCF transmission presents an

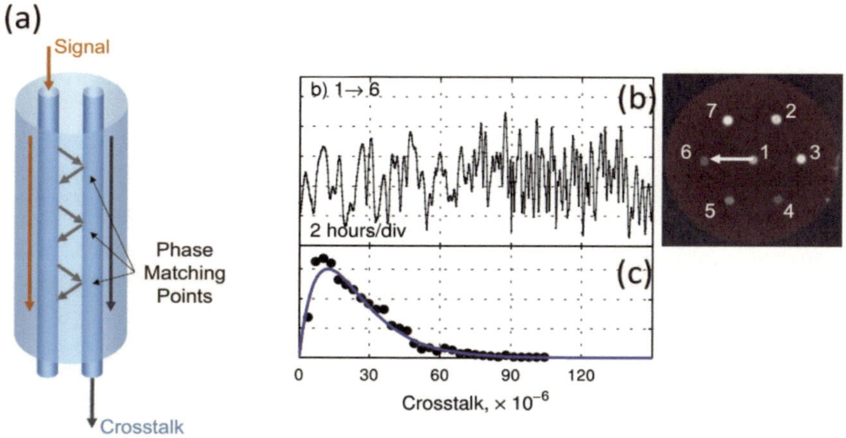

Figure 14.
(a) Mechanism of inter-core crosstalk, (b) time evolution of XT, (c) histogram of XT [12].

immense difficulty because to crosstalk, which may decrease the quality of optical data, caused by unintentional coupling between cores running in the same direction and wavelength, as shown in **Figure 14**.

Transmission is hampered by crosstalk in an MCF-based optical network, which may be reduced via power and mode coupling. So cladding widths vary. The cladding's strength decreases with increasing diameter. Each of these parameters must be changed. The MCF inter-core crosstalk is now calculated using coupled-mode and coupled power theories. First, MCF systems must agree on allowable crosstalk per length. Modern coherent optical communication systems have MCF crosstalk requirements regardless of transmission distance. Tolerable crosstalk has penalties for capacity, reach, universality transponder installation, and system link implementation [48, 49].

Ratio-reach trade-off of optical transponders capable of fine-tuning their modulation format to the channel circumstances through methods like probabilistic constellation shaping (PCS) for Nyquist pulses,

$$SE = 2.\log_2\left[1 + \frac{1}{\eta_{TRX}}\frac{P_S}{\left(\eta_L P_{ASE} + \chi P_S^3 + kP_s\right)}\right] \tag{1}$$

where SNR $= \frac{P_S}{\left(\eta_L P_{ASE}+\chi P_S^3+kP_s\right)}$.

Ps is the per-channel (dual-polarization) signal launch power, and PASE is the per-channel (dual-polarization) amplified spontaneous emission power. For example, non-perfect amplification causes noise enhancement, beginning with PASE as the ASE from ideal distributed amplification. The parameter represents nonlinear interference noise (NLIN) and is calculated utilizing [50] formalisms. It indicates the average XT power due to other signals co-propagating at the same wavelength in different MCF cores. In the low coupling regime studied here, XT may be represented as AWGN, k increases linearly with distance, and interactions between XT and fiber nonlinearities can be disregarded.

MCF optical network crosstalk research for spectrum and fiber core allocations are many. As a consequence, to reduce crosstalk, all of the methods suggested reduced network capacity. A nearby core is already transmitting data on the same wavelength, therefore they do not send data on it. Recent work in [51] shows that optical signal counter-propagation across MCF cores may decrease crosstalk.

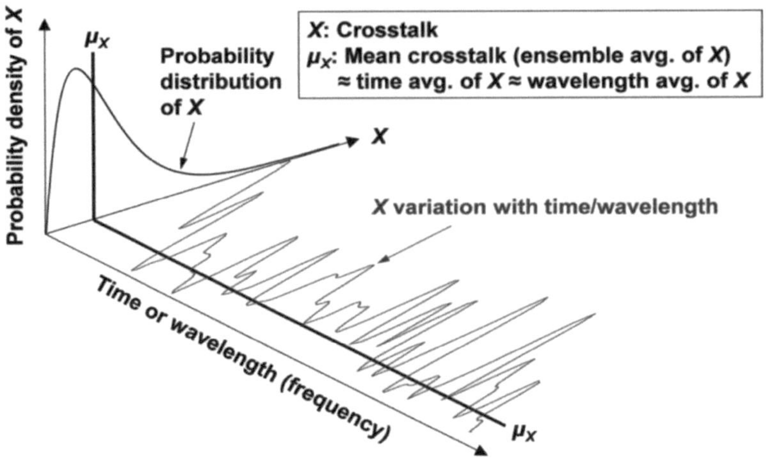

Figure 15.
Schematic diagram explaining stochastic behaviors and statistical parameters of XT in MCF [28].

Figure 15 shows XT's stochastic nature. The ensemble average X is characterized by the unpredictability of its behavior. The values of X are present in the MCF literature, but not explicitly stated. The instantaneous frequency of light is constant while the spectrum of the signal light is flat. Components of XT components that are sufficiently broad act as Gaussian noise, also called ASE noise, or nonlinear interference noise.

5.2 Non linearity

As a result of the numerous nonlinear distortions, the traditional single mode fiber properties are harmed. Many kinds of SDM fibers have been proposed. In all optical transmission and network scenarios, no single SDM fiber seems to be better than parallel SMF or ribbon fiber. To mention a few, connected core MCFs (CC-MCFs) have shown to be more resistant to non-linearity, resulting in a longer transmission distance as well as processing and amplification benefits. However, FMFs or FM-MCFs have more spatial channels per cladding diameter, making them more suited for short-distance, high-capacity connections [12, 17].

5.2.1 Self phase modulation (SPM)

The intensity dependence of the refractive index is the main source of SPM (i.e. optical Kerr effect). The change in refractive index as the signal passes through the fiber. Positive and negative refractive index gradients have leading and trailing edges. The total load at the user end of a PoF connection is dictated by the PV cell's conversion efficiency (detector). Eq. (2) states the mathematical connection, Eq. (3).

$$P_{load} = P_{in} * \eta_{pv} \tag{2}$$

$$P_{in} = \frac{P_{load}}{\eta_{pv}} \tag{3}$$

The nonlinear phase change due the SPM is given by the Eq. (**4**).

$$\varphi_{nl} = k_{nl} * P_{in} * L_{eff} \tag{4}$$

Where k_{nl} denotes the nonlinear component of the propagation constant and L_{eff} denotes the effective length. The following phase equation may be recast in terms of P_{load}, demonstrating that the nonlinear phase distortion in Eq is caused by load power (5).

$$\varphi_{nl} = k_{nl} * \frac{P_{load}}{\eta_{pv}} * L_{eff} \tag{5}$$

5.2.2 Cross phase modulation (XPM)

It is a nonlinear optical phenomenon produced by intensity changes in refractive index. XPM is guided by SPM since both rely on the refractive index and intensity of separate transmission pulses. Asymmetric spectrum broadening and signal distortion are caused by power and refractive index changes. The effective refractive index is given by Eq. (6).

$$\eta eff = \eta\,l + \eta \mathrm{nl}\left(\frac{P_{in}}{A_{eff}}\right) \tag{6}$$

This is the linear component of the refractive index profile. Similarly, as seen in Eq., the propagating constant is defined as linear and nonlinear (7)

$$k_{eff} = k_l + k_{nl} * A_{eff} \tag{7}$$

The effective refractive index and propagation constant are proportional to the effective area. It is possible that the number of cores required for high power applications has a substantial effect on the link's nonlinear distortion. Also in Eq., the core multiplicity factor [21] and N-number of core and D-cladding diameters define the A_{eff} (8)

$$A_{eff} = \left[CMF * \left(\frac{D}{2}\right)^2 * \pi\right] / N \tag{8}$$

Thus the Eq. (7) can be re written

$$\eta_{eff} = \eta_l + \eta_{nl} \frac{N}{CMF * (D/2)^2 * \pi} \tag{9}$$

According to Eq. (9), the effective refractive index (cause of XPM) is likewise affected by the number of cores in MCF.

5.2.3 Stimulated Raman scattering (SRS)

MCF has a large doped area where several optical beams may propagate. SRS is created when nonlinear acoustic vibrations interact with optical photons. The overlapping of the signal and pump electric fields at different excitation settings determines SRS efficacy. The effect of SRS for MCF for PoF connection has not been investigated. Because A_{eff} and L_{eff} influence threshold power, the number of cores, cladding diameter, and input pump power impact output power. Long-distance transmission weakens power signals, and optical beams' frequency changes downstream, producing signal loss [52].

5.2.4 Stimulated Brillouin scattering (SBS)

The performance of every optical link is affected by scattering. The number of cores improves the fiber's high power transmission capacity while decreasing the back scattered photon power, which influences the medium's nonlinearity and the acoustic photon. Within the core region, both weakly and strongly connected cores may be linear, triangular, rectangular matrix, tightly spaced hexagonal, or any other symmetric or asymmetric structure. The small core pitch type fiber has the greatest crosstalk and possible photo interaction. A PoF connection's maximal optical power transmission is limited by this interaction. Since the large numerical aperture (NA) is responsible for beam diffraction, the number of cores determines the SBS threshold. SBS changes depending on the medium's characteristics (homogeneous or birefringent) and the optical source. The thermally generated photon field affects the spectral breadth. The temperature of the fiber and its surroundings induce heat dispersion. The strain produced by internal heat may damage the fiber, reducing the output optical power. SBS has a lower effect on MCF than single mode fiber [53].

5.2.5 Optical pulse compression

MCF uses pulse compression and combination extensively. All MCF cores combine the injected optical signal. Structure and density determine signal compression. The MCF's nonlinearity produces self-focusing, anomalous dispersion, and wave collapse at high power levels. Due to the constant distance between MCF cores, the spatial non-uniformity of coupling is very important. The coupling coefficient, which determines Gaussian statistics, fluctuates with distance. Inhomogeneity in coupling causes phase mismatches and pulse delays that require special care [54].

5.2.6 Capacity wastage

If network capacity is bidirectional, overusing data centers wastes considerable capacity. To minimize MCF network effects, we asymmetrically distribute the fiber cores. It minimizes inter-core interference and allows for varying the amount of fiber cores on each side of a fiber connection. This minimizes network capacity wastage owing to mismatched bidirectional traffic demand. The suggested approach is tested on the MCF optical network's routing, spectrum, and core assignment (RSCA) problems. Two ILP models and a graph-based heuristic method are suggested to improve network spectrum usage [51].

6. Conclusion and future direction

Over the last decade, it has become apparent that MCF technologies are the only viable solution to the optical network's "capacity crunch" and other issues. Due to fiber nonlinearity, which limits growing transmission power and amplifier bandwidth. MCF should be operating by 2025. SDM's endurance and demand for telecommunications services must be shown. MCF just exceeded SSMF's maximum capacity. Increased capacity, dependability, and cost-effectiveness are required to allow broad use. New possibilities in multi-mode, spatial coding, and efficient DSP are anticipated to improve the performance of next-generation optical communication systems. This chapter examines the realities of multi-core fiber-based SDM optical wiring. The most common SDM fiber is the UC-MCF. SDM fibers use MIMO DSP to cope with modal XT.

Bidirectional traffic demand asymmetry is growing, leading to substantial capacity waste while building and running an optical transport network. Asymmetric and counter-propagating MCF fiber core allocation is advised for MCF optical networks. Assigning a flexible number of fiber cores in opposing directions to reduce network capacity waste owing to asymmetric traffic demand.

Inter-core crosstalk and traffic demand imbalance are significant factors in MCF optical network design. This network's design reduces inter-core crosstalk and capacity waste owing to bidirectional traffic demand imbalance.

Assemblies and PIC fabrication procedures are all part of SDM. When light couples the fibers, the cores converge and link the PICs. It includes extending the cores and attaching them to the PIC entrances through photonic wire bonding. Complicated handling and fusing are needed, but time-control introduces propagation delays.

Power over fiber technique uses multicore fiber structures. This chapter examined various MCF variations. The hexagonal MCF form is recommended for high power applications. Our MCF losses were also addressed. Nonlinear distortions in MCF act differently than in SMF. Some nonlinearity compensating methods, such pre-distortion, may also help reduce the impact of such distortions. The fiber cores, modes, or a mix of both provide new difficulties and possibilities for future research.

Acknowledgements

We thank Dr. Muneer Azam for assistance with particular for comments that greatly improved the manuscript.

Author details

Muhammad Irfan Anis* and Hamdan Ali
High Performance Research Group, FEST, Iqra University, Karachi, Pakistan

*Address all correspondence to: mirfananis@iqra.edu.pk

References

[1] H. Furukawa and R. S. Luis, "Petabit-class Optical Networks Based on Spatial-Division Multiplexing Technologies," *2020 24th Int. Conf. Opt. Netw. Des. Model. ONDM 2020*, pp. 21–23, 2020, doi: 10.23919/ONDM48393.2020.9132998.

[2] Mansour A, Mesleh R, Abaza M. New challenges in wireless and free space optical communications. Opt. Lasers Eng. 2016;**89**:95-108. DOI: 10.1016/j.optlaseng.2016.03.027

[3] Al-zubaidi FMA. *Towards High Bandwidth Communication Systems : From Multi-Gbit / s over SI-POF in Home Scenarios to 5G Cellular Networks over SMF by*, no. May. 2021

[4] Y. Sagae, T. Matsui, T. Sakamoto, and K. Nakajima, "Ultra-Low Crosstalk Multi-Core Fiber with Standard 125-$µ$m Cladding Diameter for 10,000 km-Class Long-Haul Transmission," *IEICE Trans. Commun.*, p. 2019OBI0001, 2020.

[5] T. Matsui *et al.*, "Multi-Core Fiber with Standard Cladding Diameter," pp. 4–5.

[6] Matsui T, Sagae Y, Sakamoto T, Nakajima K. Design and applicability of multi-core fibers with standard cladding diameter. J. Light. Technol. 2020;**38**(21):6065-6070. DOI: 10.1109/JLT.2020.3004824

[7] Zhao Y, Hu L, Zhu R, Yu X, Wang X, Zhang J. Crosstalk-Aware Spectrum Defragmentation Based on Spectrum Compactness in Space Division Multiplexing Enabled Elastic Optical Networks with Multicore Fiber. IEEE Access. 2018;**6**(4):15346-15355. DOI: 10.1109/ACCESS.2018.2795102

[8] Kitayama KI, Diamantopoulos NP. Few-Mode Optical Fibers: Original Motivation and Recent Progress. IEEE Commun. Mag. 2017;**55**(8):163-169. DOI: 10.1109/MCOM.2017.1600876

[9] Kareem FQ et al. A Survey of Optical Fiber Communications: Challenges and Processing Time Influences. Asian J. Res. Comput. Sci. 2021;**7**(4):48-58. DOI: 10.9734/ajrcos/2021/v7i430188

[10] Van Weerdenburg J et al. 138-Tb/s Mode- and Wavelength-Multiplexed Transmission over Six-Mode Graded-Index Fiber. J. Light. Technol. 2018; **36**(6):1369-1374. DOI: 10.1109/JLT.2018.2791100

[11] Y. Awaji, "Review of space-division multiplexing technologies in optical communications," *IEICE Trans. Commun.*, no. 1, 2019, doi: 10.1587/transcom.2017EBI0002.

[12] Awaji Y et al. High-capacity transmission over multi-core fibers. Opt. Fiber Technol. 2017;**35**:100-107. DOI: 10.1016/j.yofte.2016.09.008

[13] Y. Xie, L. Pei, J. Sun, J. Zheng, T. Ning, and J. Li, "Optimal design of a bend-insensitive heterogeneous MCF with differential inner-cladding structure and identical cores," *Opt. Fiber Technol.*, vol. 53, no. March, p. 102001, 2019, doi: 10.1016/j.yofte.2019.102001.

[14] Y. Xie, L. Pei, J. Zheng, Q. Zhao, T. Ning, and J. Li, "Low-DMD and low-crosstalk few-mode multi-core fiber with air-trench/holes assisted graded-index profile," *Opt. Commun.*, vol. 474, no. June, p. 126155, 2020, doi: 10.1016/j.optcom.2020.126155.

[15] C. Antonelli, G. Riccardi, T. Hayashi, and A. Mecozzi, "Role of polarization-mode coupling in the crosstalk between cores of weakly-coupled multi-core fibers," *Opt. Express*, vol. 28, no. 9, p. 12847, 2020, doi: 10.1364/oe.391092.

[16] H. Yuan *et al.*, "Experimental Analysis on Variations and Accuracy of Crosstalk in Trench-Assisted Multi-core Fibers," pp. 1–14, 2020, [Online]. Available: http://arxiv.org/abs/2008.08034.

[17] A. K. Vyas, "Analysis of different structure and nonlinear distortion of multicore fiber for power over fiber applications," *Optik (Stuttg).*, vol. 168, pp. 184–191, 2018, doi: 10.1016/j.ijleo.2018.04.106.

[18] Shibahara K, Mizuno T, Lee D, Miyamoto Y. Advanced MIMO Signal Processing Techniques Enabling Long-Haul Dense SDM Transmissions. J. Light. Technol. 2018;**36**(2):336-348. DOI: 10.1109/JLT.2017.2764928

[19] I. Morita, K. Igarashi, H. Takahashi, T. Tsuritani, and M. Suzuki, "Trans-oceanic class ultra-long-haul transmission using multi-core fiber," *Opt. Express*, vol. 22, no. 26, p. 31761, 2014, doi: 10.1364/oe.22.031761.

[20] Mendinueta JMD, Shinada S, Hirota Y, Luis RS, Furukawa H, Wada N. Converged inter/intradata center optical network with packet super-channels and 83.33 Tb/s/port. J. Light. Technol. 2019;**37**(2):571-578. DOI: 10.1109/JLT.2018.2877815

[21] M. Jiang, C. Chen, B. Zhu, and F. Hu, "MIMO-free WDM–MDM bidirectional transmission over OM3 MMF," *Opt. Commun.*, vol. 473, no. April, p. 125988, 2020, doi: 10.1016/j.optcom.2020.125988.

[22] K. Nakajima, T. Matsui, K. Saito, T. Sakamoto, and N. Araki, "Space division multiplexing technology: Next generation optical communication strategy," in *2016 ITU Kaleidoscope: ICTs for a Sustainable World (ITU WT)*, 2016, pp. 1–7.

[23] Kaur G, Kaur G, Sharma S. Performance optimization of broadband communication system using hybrid parametric amplifier. Int. J. Appl. Eng. Res. 2017;**12**(14):4484-4490

[24] Nakajima K, Matsui T, Saito K, Sakamoto T, Araki N. Multi-core fiber technology: next generation optical communication strategy. IEEE Commun. Stand. Mag. 2017;**1**(3):38-45

[25] Sakamoto T et al. Spatial Density and Splicing Characteristic Optimized Few-Mode Multi-Core Fiber. J. Light. Technol. 2020;**38**(16):4490-4496. DOI: 10.1109/JLT.2020.2987351

[26] T. Hayashi *et al.*, "Field-Deployed Multi-Core Fiber Testbed," *OECC/PSC 2019 - 24th Optoelectron. Commun. Conf. Conf. Photonics Switch. Comput. 2019*, pp. 1–3, 2019, doi: 10.23919/PS.2019.8818058.

[27] B. J. Puttnam *et al.*, "Characteristics of homogeneous multi-core fibers for SDM transmission," *APL Photonics*, vol. 4, no. 2, 2019, doi: 10.1063/1.5048537.

[28] T. Hayashi and T. Nakanishi, "Multi-core optical fibers for the next-generation communications," *SEI Tech. Rev.*, no. 86, pp. 23–28, 2018.

[29] T. Hayashi, T. Taru, O. Shimakawa, T. Sasaki, and E. Sasaoka, "Uncoupled multi-core fiber enhancing signal-to-noise ratio," *Opt. Express*, vol. 20, no. 26, p. B94, 2012, doi: 10.1364/oe.20.000b94.

[30] Puttnam BJ et al. 0.61 Pb/s S, C, and L-Band Transmission in a 125μm Diameter 4-Core Fiber Using a Single Wideband Comb Source. J. Light. Technol. 2021;**39**(4):1027-1032. DOI: 10.1109/JLT.2020.2990987

[31] Cho J et al. Trans-Atlantic Field Trial Using High Spectral Efficiency Probabilistically Shaped 64-QAM and Single-Carrier Real-Time 250-Gb/s 16-QAM. J. Light. Technol. 2018;**36**(1): 103-113. DOI: 10.1109/JLT.2017.2776840

[32] Peng G-D. *Handbook of Optical Fibers*. Springer; 2019

[33] Xie X, Tu J, Zhou X, Long K, Saitoh K. Design and optimization of 32-core rod/trench assisted square-lattice structured single-mode multi-core fiber. Opt. Express. 2017;**25**(5):5119-5132

[34] A. Turukhin *et al.*, "105.1 Tb/s power-efficient transmission over 14,350 km using a 12-core fiber," *2016 Opt. Fiber Commun. Conf. Exhib. OFC 2016*, pp. 105–107, 2016, doi: 10.1364/ofc.2016.th4c.1.

[35] Z. Luo, S. Yin, L. Jiang, L. Zhao, and S. Huang, "Routing, Spectrum and Core Assignment based on Auxiliary Matrix in the Intra Data Center Networks using Multi-Core Fibers with Super Channel," in *Asia Communications and Photonics Conference/International Conference on Information Photonics and Optical Communications 2020 (ACP/IPOC)*, 2020, p. M4A.279, doi: 10.1364/ACPC.2020.M4A.279.

[36] J. M. D. Mendinueta, S. Shinada, Y. Hirota, H. Furukawa, and N. Wada, "High-Capacity Super-Channel-Enabled Multi-Core Fiber Optical Switching System for Converged Inter/Intra Data Center and Edge Optical Networks," *IEEE J. Sel. Top. Quantum Electron.*, vol. 26, no. 4, 2020, doi: 10.1109/JSTQE.2020.2969558.

[37] V. N. Korshunov, I. A. Ovchinnikova, S. S. Shavrin, N. A. Shishova, and A. Y. Tsym, "Spectral Efficiency and Information Transfer Rate over Optical Fibers in Spatial Parallelism," 2020 Syst. Signals Gener. Process. F. Board Commun., vol. 4, no. 1, pp. 0–3, 2020, doi: 10.1109/IEEECONF48371.2020.9078609.

[38] García Cortijo S. Distributed radiofrequency signal processing based on space-division multiplexing fibers. Valencia (Spain): Universitat Politècnica de València; 2020

[39] Tang F, Li Y, Shen G, Rouskas GN. Minimizing inter-core crosstalk jointly in spatial, frequency, and time domains for scheduled lightpath demands in multi-core fiber-based elastic optical network. J. Light. Technol. 2020;**38**(20): 5595-5607

[40] M. Morant, A. M. Trinidad, E. Tangdiongga, and R. Llorente, "Multi-core Fiber Technology supporting MIMO and Photonic Beamforming in 5G Multi-Antenna Systems : (Invited paper)," *2019 IEEE Int. Top. Meet. Microw. Photonics, MWP 2019*, pp. 1–4, 2019, doi: 10.1109/MWP.2019.8892041.

[41] J. Cariñe *et al.*, "Multi-core fiber integrated multi-port beam splitters for quantum information processing," *Optica*, vol. 7, no. 5, p. 542, 2020, doi: 10.1364/optica.388912.

[42] B. Annighoefer, A. Zeyher, and J. Reinhart, "Multi-core Fiber and Power-limited Optical Network Topology Optimization with MILP," 2021, [Online]. Available: http://arxiv.org/abs/2103.16981.

[43] Lee SH. Experimental study. Sch. Res. Music. 2017:81-88. DOI: 10.4324/9781315458090

[44] A. Samir, J. Ratkoceri, and B. Batagelj, "Multi-core optical fiber in a passive optical local area network," in *2018 International Conference on Innovative Trends in Computer Engineering (ITCE)*, 2018, pp. 77–82.

[45] D. Kumar and R. Ranjan, "Crosstalk Suppression using Trench-assisted Technique in 9-core Homogeneous Multi Core Fiber," *2017 14th IEEE India Counc.* Int. Conf. INDICON *2017*, pp. 6–9, 2018, doi: 10.1109/INDICON.2017.8487586.

[46] Rademacher G et al. High Capacity Transmission in a Coupled-Core Three-Core Multi-Core Fiber. J. Light. Technol. 2021;**39**(3):757-762. DOI: 10.1109/JLT.2020.3013966

[47] Lee Y, Tanaka K, Hiruma K, Nomoto E, Sugawara T, Arimoto H. Experimental demonstration of a highly reliable multicore-fiber-based optical network. IEEE Photonics Technol. Lett. 2014;**26**(6):538-540. DOI: 10.1109/LPT.2013.2296100

[48] A. Macho Ortiz, "Multi-Core Fiber and Optical Supersymmetry: Theory and Applications," 2019.

[49] Gene JM, Winzer PJ. A universal specification for multicore fiber crosstalk. IEEE Photonics Technol. Lett. 2019;**31**(9):673-676. DOI: 10.1109/LPT.2019.2903717

[50] C. Antonelli, O. Golani, M. Shtaif, and A. Mecozzi, "Nonlinear interference noise in space-division multiplexed transmission through optical fibers," *Opt. Express*, vol. 25, no. 12, p. 13055, 2017, doi: 10.1364/oe.25.013055.

[51] Tang F, Yan Y, Peng L, Bose SK, Shen G. Crosstalk-Aware counter-propagating core assignment to reduce inter-core crosstalk and capacity wastage in multi-core fiber optical networks. J. Light. Technol. 2019; **37**(19):5010-5027. DOI: 10.1109/JLT.2019.2927025

[52] Ceballos-Herrera DE, Gutierrez-Castrejon R, Alvarez-Chavez JA. Stimulated raman scattering and four-wave mixing effects on crosstalk of multicore fibers. IEEE Photonics Technol. Lett. 2018;**30**(1):63-66. DOI: 10.1109/LPT.2017.2774501

[53] Z. Zhao, M. A. Soto, M. Tang, and L. Thévenaz, "Demonstration of distributed shape sensing based on Brillouin scattering in multi-core fibers," *25th Int.* Conf. Opt. Fiber Sensors, vol. 10323, p. 1032393, 2017, doi: 10.1117/12.2267486.

[54] A. M. Rubenchik, I. S. Chekhovskoy, M. P. Fedoruk, O. V. Shtyrina, and S. K. Turitsyn, "Nonlinear pulse combining and pulse compression in multi-core fibers," *Opt. Lett.*, vol. 40, no. 5, p. 721, 2015, doi: 10.1364/ol.40.000721.

Chapter 2

Functional Tapered Fiber Devices Using Polymeric Coatings

Oscar González-Cortez, Rodolfo A. Carrillo-Betancourt, Juan Hernández-Cordero and Amado M. Velázquez-Benítez

Abstract

A wide variety of fiber devices can be created by adding special coatings on tapered sections of optical fibers. In this work we present the fundamentals for the fabrication of tapered optical fibers coated with functional polymers. The required aspects of light propagation in tapered sections of optical fibers are introduced and the relevant parameters enabling light interaction with external media are discussed. A special case of interest is the addition of polymeric coatings with prescribed thicknesses in the tapered sections allowing for adjusting the light propagation features. We assess the use of liquid polymer coatings with varying thicknesses along the taper profile that can be tailored for tuning the transmission features of the devices. Hence, we introduce a methodology for obtaining coatings with predefined geometries whose optical properties will depend on the polymer functionality. As demonstrated with numerical simulations, the use of functional polymer coatings in tapered optical fibers allows for obtaining a wide variety of functionalities. Thus, controlled polymer coating deposition may provide a simple means to fabricate fiber devices with adjustable transmission characteristics.

Keywords: Tapered fibers, thin coatings, polymers, evanescent wave

1. Introduction

Optical fiber-based devices and sensors have been widely used in many fields of science and technology for different applications. In most applications, the transmission of light through in-line fiber devices is modified by an external perturbation and this can be quantified through variations in one or more characteristic features of the guided optical wave. Although extrinsic interaction of the light with the surrounding media is possible, the use of the evanescent portion of the guided wave offers some advantages. Exposure of the optical wave to the surrounding media in specific sections of the optical fibers is usually done by two alternatives [1–5]: removing the cladding material, or upon tapering a section of the optical fiber. The latter is the preferred approach to expose the evanescent wave since it involves a simple and reproducible process yielding low-loss devices. Although the description and basics of tapering optical fibers have been described since early 1990's by Birks *et al.* [1], new applications for these devices have been a subject of research due to their potential use in many fields of science and technology. Some of these applications include physical and biomedical sensing [6–10], interferometry [11], excitation of surface plasmons [12, 13], atom detection [14, 15], and light coupling to micro-resonators [8, 16], just to name a few.

Tapering techniques and systems have evolved over the years resulting in optimized devices and standardized processes. As a result, tapering of optical fibers has been employed for modifying the light propagation conditions in fibers, and further allowing for the guided light to interact with other structures or materials. The latter capabilities allow for creating fiber-based devices incorporating coatings with controlled thicknesses and different optical properties. In this chapter we present the main aspects and basic guidelines for achieving a proper guidance of light through tapered devices providing also adequate interaction with the surrounding medium.

1.1 Basics of light propagation in optical fibers

Light propagation in optical fibers occurs within the fiber core (radius a) following the total internal reflection condition from Snell's law. Propagation conditions are defined by the physical characteristics of the core and cladding sections of the fiber: refractive indices and diameters. Due to the cylindrical geometry of the optical fibers and the small refractive index difference between the core and cladding materials, the propagating modes are obtained in a cylindrical coordinate system (r, φ, ϕ) and in terms of Bessel equations [17, 18]. Rigorous analysis and proper boundary conditions are used to obtain mathematical expressions describing the components of the electric and magnetic fields for the core ($r \leq a$) and the cladding ($r \geq a$) regions. The combination of these propagating fields allows for an alternative description in terms of linearly polarized modes, denoted as LP_{lm}, in which l and m are respectively the axial and radial indices. As an example, the electric field for the set of LP_{0m} modes can be shown to be given by:

$$E_{LP_{0m}} = \begin{cases} E_0 J_0\left(\frac{u\,r}{a}\right), & r \leq a \\ E_0 \dfrac{J_0(u)}{K_0(w)} K_0\left(\frac{w\,r}{a}\right), & r > a \end{cases}, \tag{1}$$

where E_0 is the field amplitude, J_0 is the zeroth-order Bessel function of the first kind, and K_0 is the zeroth-order modified Bessel function of the second kind. The parameters u and w are respectively the normalized propagation and attenuation constants, defined as:

$$u = a\sqrt{k^2 n_{core}^2 - \beta^2}, \quad w = a\sqrt{\beta^2 - k^2 n_{cladd}^2}, \tag{2}$$

involving the propagation constant of the guided wave (β), the wave number (k) and the refractive indices of the core (n_{core}) and the cladding (n_{cladd}). The description in terms of the LP_{lm} modes is very useful as these can be experimentally observed as intensity patterns (I_{lm}) whose mathematical representations are [17]:

$$I_{lm} = \begin{cases} I_0 J_l^2\left(\frac{u\,r}{a}\right) sin^2(l\phi), & r \leq a \\ I_0 \left(\frac{J_l(u)}{K_l(w)}\right)^2 K_l^2\left(\frac{w\,r}{a}\right) sin^2(l\phi), & r > a \end{cases}. \tag{3}$$

Another important parameter to assess the modal features of an optical fiber is the normalized frequency or V number, defined as:

$$V = \sqrt{u^2 - w^2} = \frac{2\pi a \sqrt{n_{core}^2 - n_{cladd}^2}}{\lambda}. \tag{4}$$

The normalized frequency is typically used as an indicator of how many modes are supported by the fiber. It can be shown that standard single mode fibers (SMFs) supporting only the fundamental (LP_{01}) mode have a $V < 2.405$. Hence, this condition is typically used as an indicator of single-mode propagation in an optical fiber.

1.2 Light propagation in tapered optical fibers

Tapered optical fibers are fabricated such that the physical dimensions of the core and cladding are reduced, thereby modifying the light propagation conditions [8, 19, 20]. The implication of this reduction in the physical dimensions of the fiber is a decrease in the effective refractive index of the core and thus a change in the light confinement. An effect caused by the core reduction is the compression of the light inside the core, although the evanescent wave increases in magnitude. However, there is a physical limit for the core reduction since the light cannot be effectively guided inside this region under certain conditions. After reaching this "guiding threshold" in the dimensions of the core, light escapes to the cladding material, which acts as the new core of the optical fiber while the surrounding media becomes the new cladding. As a consequence, the evanescent wave is effectively exposed to the external medium, granting the possibility of light interaction with different materials or structures.

Typically, tapered optical fibers are segmented in three sections for their analysis [1]: *the non-tapered segment of the optical fiber*, the *transition* and the *waist sections*, as depicted in **Figure 1a**. The waist section of the fiber is where the fiber is tapered down to its final diameter after which it remains constant across all its length. Usually, the interaction or coupling to other structures or materials takes place at this section. The transition sections of the fiber are those at which the diameter goes from the original diameter to the tapered section, and vice versa. Although the transition regions are commonly not used for sensing or coupling, they are significantly important as disturbances or non-desired alterations to light can occur in these sections. Such effects appear when the transition between diameters is not smooth or non-adiabatic.

Adiabaticity criteria comes from the tapering angle, which must be small enough for the fundamental mode to smoothly propagate across all the sections with minimal power loss [19]. Any sudden change in the dimensions or geometry of the fiber will provoke imminent light leaking from the core. This is avoided by performing the tapering process at a very slow rate. Considering the case of tapering a SMF, an adequate transition should maintain the fundamental mode through all the sections of the tapered fiber with negligible losses. Conversely, a non-adiabatic tapering

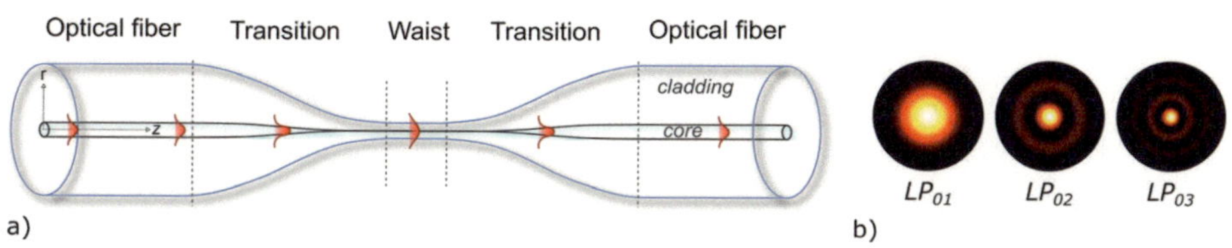

Figure 1.
(a) Tapered optical fiber profile indicating the three zones: non-tapered, transition and waist. (b) Intensity distributions of the LP_{om} *modes excited in the transition section and propagating in the waist of the taper.*

process will produce a "leakage" of light into the cladding exciting higher order modes.

The parameter used to define the adiabaticity of the tapered fiber is the *local length-scale* of the taper (z_t), resulting from the ratio of the core radius (r) over the tapering angle (Ω), expressed as $z_t = \frac{r}{\Omega}$. Notice that both variables (r and Ω) vary with length, i.e., $r = r(z)$ and $\Omega = \Omega(z)$ [19]. To achieve adiabaticity, the length of the taper must be larger than the length required to couple the fundamental mode to the predominant cladding mode. This coupling length is given by the beat length (z_b), which in turn depends on the propagation constant, $\beta = kn_{eff}$, with $k = \frac{2\pi}{\lambda}$. Thus, this variable is given by the expression [19]:

$$z_b = \frac{2\pi}{\beta_1 - \beta_{CM}}, \tag{5}$$

where β_1 is the propagation constant of the fundamental LP_{01} mode, and β_{CM} corresponds to the propagation constant of the excited cladding mode. In an axially symmetric tapered fiber, the fundamental mode will only couple to azimuthally symmetric higher order modes (i.e., the LP_{0m} modes) supported by the cladding structure (see **Figure 1b**).

To illustrate the phenomena related to tapering optical fibers, we will use as an example a step-index standard single-mode silica fiber with core radius of 4.1 μm, cladding radius of 62.5 μm and refractive index difference $\Delta n = 0.061$. These are typical dimensions for the standard telecommunication fibers that are commonly used for the fabrication of tapered fibers for evanescent coupling of light to other photonic devices. Upon considering an adiabatic tapering process, firstly can be noticed that by reducing the core radius the V number is also modified. This implies that the effective refractive index (n_{eff}) of the waveguide will be modified as well and thus the propagation constant. The solutions in terms of the Bessel approximations yield the values of the propagation constants (β) resulting from these changes in geometry [17, 18]. **Figure 2** shows results from numerical calculations illustrating that a reduction in the propagation constant is obtained as the core radius decreases. Notice also that for $V < 1$, a sudden decrease in β occurs indicating that light cannot be longer confined in the core and hence it couples to the cladding material. Such

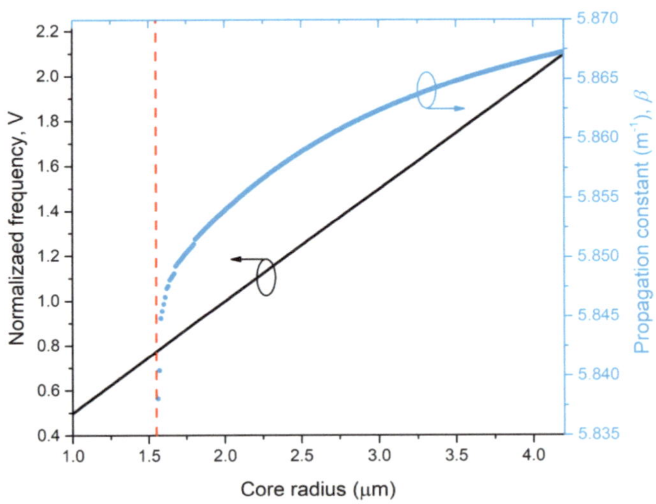

Figure 2.
Variation of the normalized frequency (V) and propagation constant (β) for different core radii corresponding to different tapering ratios. The dashed line shows the "guiding threshold" of the fiber core.

phenomenon implies that the whole material of the optical fiber, core and cladding, acts together as the new core of the waveguide, while the surrounding medium serves as the new cladding. The first reports on this phenomenon mention values for this guiding threshold V_{TH} near 0.8 [8, 19] to achieve this condition. Our numerical results indicate that light confinement in the fiber core is sustained until $V_{TH} \approx 0.77$, which corresponds to an approximate core radius of 1.55 μm, as illustrated in **Figure 2**.

The intensity distribution of the fundamental mode is modified as the fiber is tapered. For core diameters above the guiding threshold, a tighter confinement of light within the core is achieved. However, the amount of light in the evanescent portion of the wave increases and so its extension inside the cladding. Once the guiding threshold is surpassed, light confinement shifts from the core to the cladding, distributing across the entire waist material. This effect is illustrated in **Figure 3**, showing light confinement at different tapering ratios. Instantly, as the guided light is exposed to the external media, light confinement and propagation are determined by the surrounding materials. At this stage, interaction with the surrounding media occurs via the evanescent wave. Nonetheless, the amount of light exposed to the external media can be tailored by defining the final waist diameter. For a fiber surrounded by air, due to the large refractive index difference with respect to the silica, the V number substantially increases allowing to reduce the fiber core down to very small diameters. Subwavelength waist diameters have been reported for tapered fiber devices in sensing and other applications [8, 15, 21].

The wavelength of light is also relevant in this phenomenon as in every waveguiding structure. Light confinement in the fundamental mode will change depending on the wavelength, showing a larger amount of evanescent wave for

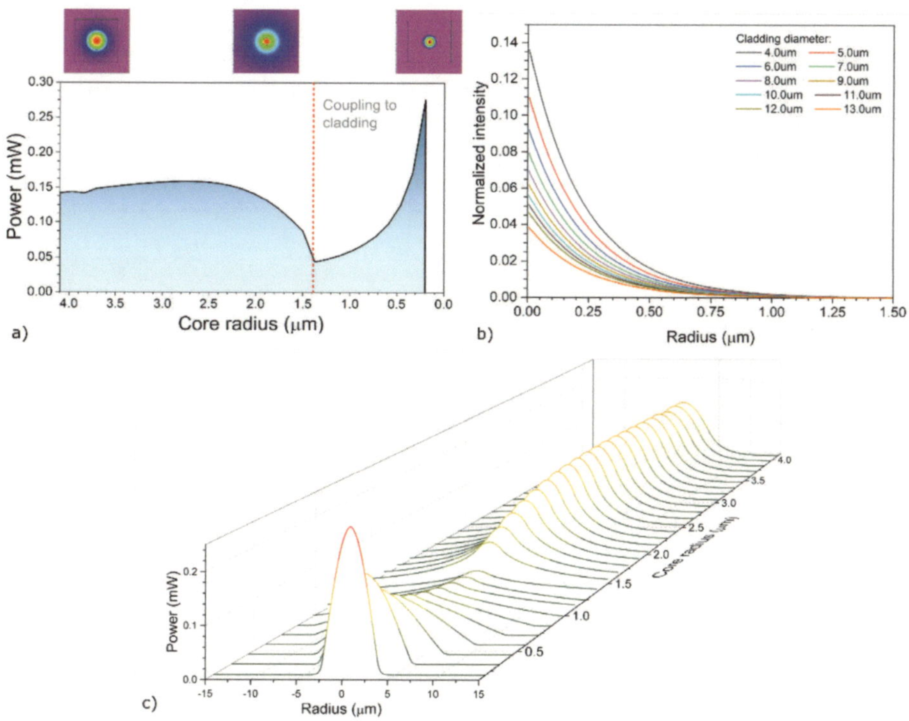

Figure 3.
Light confinement in tapered optical fibers. Fundamental mode power evolution at different core radii corresponding to different tapering ratios: (a) peak power and (b) power distribution. (c) Evanescent wave extension for different tapering ratios.

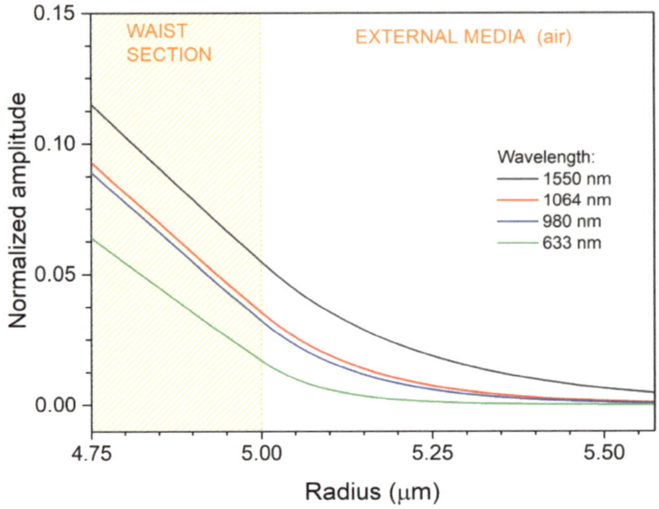

Figure 4.
Wavelength dependence of the penetration of the evanescent wave of the fundamental mode to external media at the waist of a tapered optical fiber.

longer wavelengths. Assuming an adiabatically tapered SMF with a 5 μm waist radius with air as the surrounding medium, the extension of the evanescent wave will present a significant increase as illustrated in **Figure 4**. Hence, for a given tapered device with specific dimensions, wavelength selection is essential as this will change the exposure and hence the interaction of the evanescent wave with the external medium.

Light exposure through evanescent wave is useful for the creation of functional devices by means of adding different materials as coatings. The amount of evanescent wave exposed to the surrounding media determines the range of coating thicknesses that can be effectively used for efficient interaction depending on the materials as well. The fundamentals of fiber coating and main aspects to consider for the creation of devices using diverse materials as coatings are discussed in the following sections.

2. Fiber coating

There are multiple reports using diverse techniques for coating optical fibers with tapered sections. We will focus on a method based on the deposition of liquid materials over specific sections of optical fibers since these are convenient for post-processing fibers. We consider in particular the use of liquid polymers as coating materials, since they offer a wide variety of functionalities. Polymers have been used for recoating fiber devices such as Bragg gratings, amplifiers or splices, just to mention the most important. Various works have demonstrated the effectiveness of using this technique for diverse purposes [22–25], and recoating systems to perform such task have been readily reported [26].

The process for deposition of controlled layer coatings on optical fibers using liquid materials, such as polymers, are suitably described by the wire coating technique [27–29]. This involves the immersion of the fiber into the liquid and retrieve it at a prescribed speed to obtain a desired thickness, as illustrated in **Figure 5**. The main factors involved in this process include the characteristics of the liquid such as viscosity (η) and surface tension (γ), and the velocity (v) at which the process takes

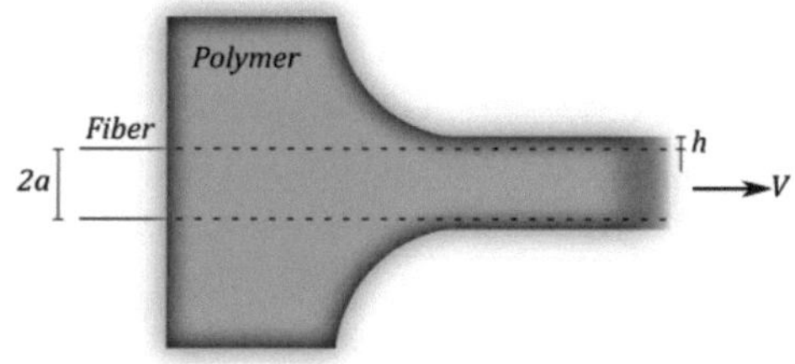

Figure 5.
Wire coating technique scheme. A layer of coating material (thickness h) is left on the surface of the wire (radius a) by pulling the coating material at a constant velocity (v) through a liquid container.

place. The ratio of the viscosity to the surface tension, multiplied by the velocity of the coating process, is defined as the capillary number, which allows to compare the behavior of different liquids. This ratio is a dimensionless number defined explicitly as $Ca = {\eta v}/{\gamma}$. For liquids with $Ca < < 1$, the resulting coating thickness (h) deposited on a fiber of radius (a) can be obtained as [29]:

$$h = 1.34aCa^{2/3}. \tag{6}$$

This expression holds only for $Ca < 1$ since the fluid dynamics are different for larger values of the capillary number. Nonetheless, proper adjustments to this theory can be made yielding a correction factor for the thickness [30]. This wire coating theory has been largely investigated and described in the field of fluidics describing in detail multiple scenarios and materials.

Given the nature of the liquids, an additional factor to consider for their use as fiber coatings is the Rayleigh-Plateau instability [31]. This effect leads to the breakdown of the uniform layer into a periodic array of droplets. Such behavior will always occur at characteristic time (t_O) given by:

$$t_O = 12\frac{\eta(r+a)^4}{\gamma h^3}. \tag{7}$$

This is an important effect that must be considered when applying liquid coatings because it indicates the maximum time in which the liquid layer will remain with a uniform thickness. Therefore, the coating must be solidified before this characteristic time in order to preserve its shape. Hence, different polymeric materials cured by means of chemical, thermal or photo-active processes must account for this breakdown time.

3. Coatings on tapered optical fibers

In most cases, the coatings on the tapered sections of the fiber are sought to be with uniform thickness in order to obtain interaction with the evanescent wave. Ideally, only the waist section of the taper should be coated, as indicated in the green region in **Figure 6a**. However, there are two aspects that must be also considered when designing the coating: the actual exposition of the evanescent wave and the coating process itself. In practice, coating only the waist section of the tapered fiber might represent a challenging task since the liquid to be deposited is usually contained within a reservoir with prescribed dimensions that do not match the length of the waist. Extraction of the fiber from the reservoir can also be a potential factor to

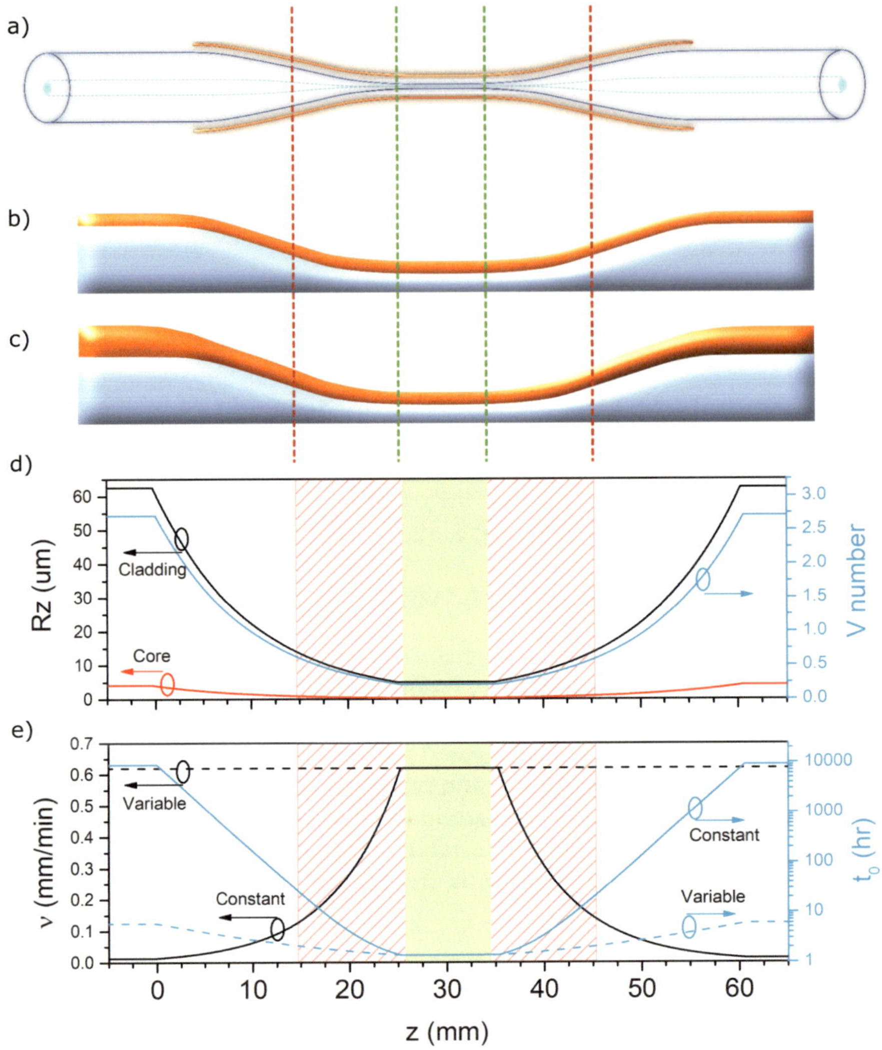

Figure 6.
Coatings on optical fibers. (a) Schematic of a coated tapered optical fiber in different regions showing the regions where coupling from the core to the cladding occurs and the modal distribution in the waist section. (b) Constant and (c) variable thickness coatings along the tapered fiber. (d) Variation of the normalized frequency (V number) of the fiber core along the taper; the red regions indicate the core-cladding coupling zones and light interacts with coating; the green zone indicates the uniform waist section. (e) Coating speeds (v) *required to obtain constant and variable coating thicknesses along the taper; the plot includes the corresponding characteristic time* (t_o) *for the instabilities.*

generate a non-uniform coating; this task is usually performed upon pulling the fiber from the non-tapered section and hence liquid is also deposited in other regions of the tapered fiber. The transition zones of the tapered fiber are therefore coated as well, and this minimizes any perturbations generated by the reservoir.

Non-uniformities in the coating thickness are a potential source of perturbation that may result in light leaking in the optical fiber. Given the variable radius profile of the tapered fiber and the radius dependency for the resultant thickness given by (6), the profile of the coating can be modified depending on the velocity of the process. Then, two scenarios can result from the coating process: a uniform coating along the entire fiber obtained upon adjusting the speed (**Figure 6b**), or a variable coating layer using a constant speed during the process (**Figure 6c**). In all cases, it is also important to consider the cross-linking mechanism of the coating polymer.

Once the coating is deposited, it might require a longer period of time for achieving solidification.

To exemplify these possibilities, we consider a tapered optical fiber coated with polydimethylsiloxane elastomer (PDMS, Sylgard 184 from Dow), which has a $\eta = 3.5[Pa \bullet s]$ and $\gamma = 0.0198[N \bullet m^{-1}]$ [26, 32]. For the SMF tapered down to a waist radius of 5 μm, the lengths of the waist and transitions are 10 and 25 mm, respectively; the transitions are assumed to have an exponential profile as described in [1]. For this specific fiber light couples into the cladding, and thus evanescent wave exposition into the surrounding media occurs along the transition section (red region in **Figure 6**). The radii of the fiber core and the cladding are compared with the modification of the normalized frequency in **Figure 6d**. The nature of the curing or solidifying process of the polymer is crucial and should be considered due to the possible appearance of instabilities. To avoid this, both the coating and curing processes needs to be shorter than the characteristic time (t_O) in the waist section.

PDMS is a polymer cured by heating, requiring from several minutes to hours for complete solidification, depending on the temperature of the curing process. For achieving a 200 nm thick uniform PDMS coating at the waist section, a speed coating process of around $\nu = 0.62\left[mm \bullet min^{-1}\right]$ is required, and the critical time is $t_0 = 1.4[hr]$. The conditions required for generating a coating with constant or variable thickness layers are analyzed in **Figure 6e**. For obtaining a coating with constant thickness (solid line), the required coating speed increases inversely with the fiber diameter. Although the instabilities mostly appear at larger times along the tapered fiber, the total duration of the process will be in general longer. In contrast, the variable coating (dashed line) implicates overall shorter times for instabilities to appear, and the process must be performed much faster.

Depending on the application, fiber coatings can be designed to optimize the interaction of the evanescent wave with the surrounding media. This means that the thickness of the coating can be extended to completely enclose the evanescent wave, or simply to favor some interaction and leave a remaining portion of light still interacting with the external media. Having this in mind, one must also consider the refractive index of the material: while low refractive index materials will require thinner coatings to completely isolate the evanescent wave, high refractive index

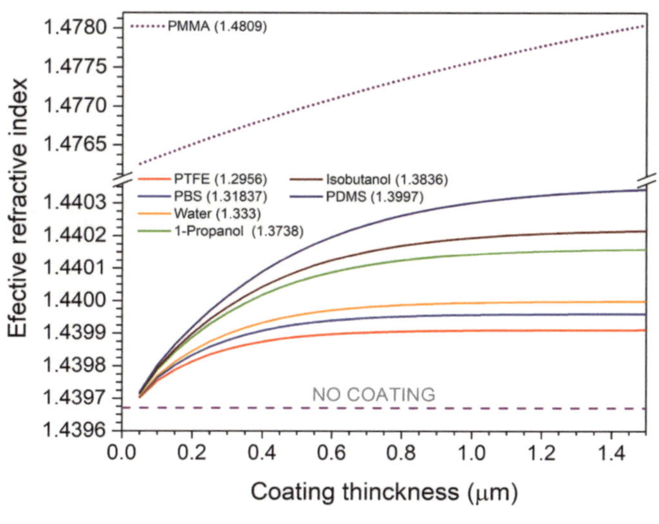

Figure 7.
Effective refractive index at the waist section of the fiber as a function of the coating thickness for materials commonly used for taper coating and for sensing applications: PTFE [33], PBS [34], water [34], 1-propanol [35], isobutanol [35] and PDMS [26, 32].

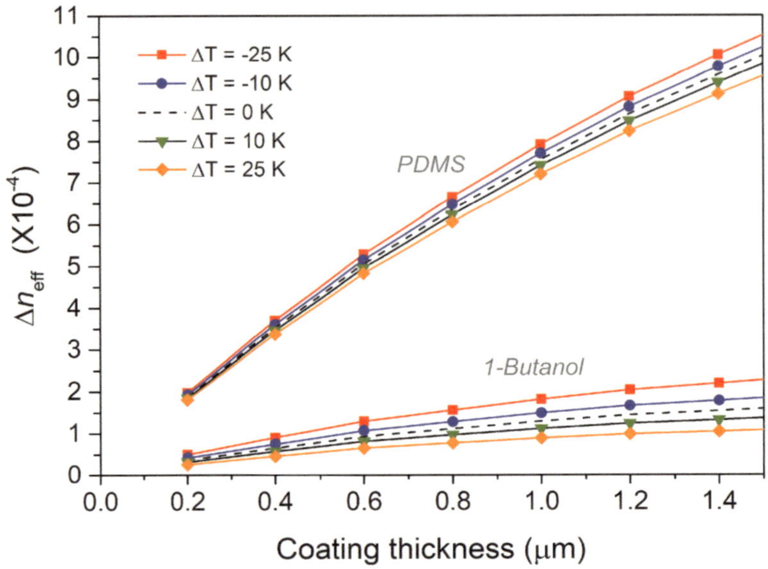

Figure 8.
Temperature dependence of the LP_{01} mode effective refractive index (n_{eff}) for different thicknesses of PDMS coatings.

materials will allow to use thicker coatings before isolating the evanescent wave, as illustrated in **Figure 7**. Of course, to have efficient and low loss light propagation in the tapered fiber, the refractive index of the coating material must be lower than the refractive index of the optical fiber. Otherwise, this will induce losses since the light propagated in the cladding will never been coupled again the core.

4. Applications of coatings for devices

Different effects can be generated in the guided wave depending on the optical properties of the coating material. For instance, the sensitivity of the tapered device to physical parameters can be tailored up to some extent. Using again as an example PDMS coatings, the effective refractive index of the fundamental mode can be adjusted with the coating thickness. It is well known that PDMS experiences changes with temperature due to its thermal expansion coefficient and to its thermo-optic coefficient ($\Delta n/_{\Delta T}[K^{-1}] = -1.8 \times 10^{-4}$) [32]. As the thickness of the PDMS coating is increased, the effective refractive index will increase its variation with temperature, as shown in **Figure 8**. Evidently, this "tunability" in sensitivity is limited, and it further depends on the thermo-optical properties of the polymer. To illustrate this, we have also included in **Figure 8** an example using 1-Butanol as the coating material [35]. Notice that in this case, lower effective indices are obtained and their variations with coating thickness are not as pronounced as obtained for PDMS. However, the sensitivity of the effective index to temperature changes for this coating is larger than that of PDMS.

5. Conclusions

We have introduced guidelines for coating tapered sections of optical fibers with liquid materials. Specifically, we undertake the subject of polymeric thin layers deposited by the wire coating technique and the interaction of the coating with the evanescent wave. Firstly, the propagation considerations to achieve light interaction with the surrounding media via evanescent wave were described. This aspect depends on the physical dimensions, refractive indices of the materials and propagating wavelength. As shown upon analyzing the wire coating technique, it is possible to define parameter for the coating process that will allow for controlling the coating thickness and avoid coating instabilities. Essentially, once the properties of the liquid polymer and the dimensions of the taper are known, the coating velocity may be calculated in order to obtain a prescribed coating thickness. Finally, light propagation for different coating thickness with different materials was discussed in terms of the influence on the effective refractive index in the waist section of the fiber. These aspects should be of interest for obtaining photonic devices with functional polymers coatings that may be useful for sensing applications.

Acknowledgements

This work was possible thanks to the funding by DGAPA-PAPIIT project TA101220. The Authors want to thank to Synopsys by the licensing of RSoft.

Conflict of interest

The authors declare no conflict of interest.

Author details

Oscar González-Cortez[1], Rodolfo A. Carrillo-Betancourt[1,2], Juan Hernández-Cordero[2] and Amado M. Velázquez-Benítez[1]*

1 Instituto de Ciencias Aplicadas y Tecnología, Universidad Nacional Autónoma de México, Mexico City, Mexico

2 Instituto de Investigaciones en Materiales, Universidad Nacional Autónoma de México, Mexico City, Mexico

*Address all correspondence to: amado.velazquez@icat.unam.mx

References

[1] Birks TA, Li YW. The shape of fiber tapers. Journal of Lightwave Technology. 1992 Apr;10(4):432-438. DOI: 10.1109/50.134196

[2] Baker, Chams, and Martin Rochette. "A generalized heat-brush approach for precise control of the waist profile in fiber tapers." Optical Materials Express 1.6 (2011): 1065-1076. DOI: 10.1364/OME.1.001065

[3] Haddock HS, Shankar PM, Mutharasan R. Fabrication of biconical tapered optical fibers using hydrofluoric acid. Materials Science and engineering: B. 2003 Jan 15;97(1):87-93. https://doi.org/10.1016/S0921-5107(02)00434-8

[4] Ko S, Lee J, Koo J, Joo BS, Gu M, Lee JH. Chemical wet etching of an optical fiber using a hydrogen fluoride-free solution for a saturable absorber based on the evanescent field interaction. Journal of Lightwave Technology. 2016 Aug 15;34(16): 3776-3784. DOI: 10.1109/JLT.2016.2583061

[5] Nikbakht, Hamed, Hamid Latifi, Mohammadreza Oraie, and Tahere Amini. "Fabrication of tapered tip fibers with a controllable cone angle using dynamical etching." Journal of Lightwave Technology 33, no. 23 (2015): 4707-4711. DOI: 10.1109/JLT.2015.2453365

[6] Villatoro, Joel, David Monzón-Hernández, and Efrain Mejía. "Fabrication and modeling of uniform-waist single-mode tapered optical fiber sensors." Applied optics 42, no. 13 (2003): 2278-2283. DOI: 10.1364/AO.42.002278

[7] Yang, Rui, Yong-Sen Yu, Cong-Cong Zhu, Yang Xue, Chao Chen, Xuan-Yu Zhang, Bao-Lin Zhang, and Hong-Bo Sun. "PDMS-coated S-tapered fiber for highly sensitive measurements of transverse load and temperature." IEEE Sensors Journal 15, no. 6 (2015): 3429-3435. DOI: 10.1109/JSEN.2015.2388490

[8] Brambilla G, Xu F, Horak P, Jung Y, Koizumi F, Sessions NP, Koukharenko E, Feng X, Murugan GS, Wilkinson JS, Richardson DJ. Optical fiber nanowires and microwires: fabrication and applications. Advances in Optics and Photonics. 2009 Jan 30;1 (1):107-161. https://doi.org/10.1364/AOP.1.000107

[9] Leung, Angela, P. Mohana Shankar, and Raj Mutharasan. "A review of fiber-optic biosensors." Sensors and Actuators B: Chemical 125, no. 2 (2007): 688-703. DOI: 10.1016/j.snb.2007.03.010

[10] Tian, Ye, Wenhui Wang, Nan Wu, Xiaotian Zou, and Xingwei Wang. "Tapered optical fiber sensor for label-free detection of biomolecules." Sensors 11, no. 4 (2011): 3780-3790. DOI: 10.3390/s110403780

[11] Yadav, T. K., R. Narayanaswamy, MH Abu Bakar, Y. Mustapha Kamil, and M. A. Mahdi. "Single mode tapered fiber-optic interferometer based refractive index sensor and its application to protein sensing." Optics Express 22, no. 19 (2014): 22802-22807. DOI: 10.1364/OE.22.022802

[12] Verma, Rajneesh K., Anuj K. Sharma, and B. D. Gupta. "Surface plasmon resonance based tapered fiber optic sensor with different taper profiles." Optics Communications 281, no. 6 (2008): 1486-1491. DOI: 10.1016/j.optcom.2007.11.007

[13] Monzón-Hernández, David, Joel Villatoro, Dimas Talavera, and Donato Luna-Moreno. "Optical-fiber surface-plasmon resonance sensor with multiple resonance peaks." Applied optics 43, no. 6 (2004): 1216-1220. DOI: 10.1364/AO.43.001216

[14] O'Shea, Danny, Christian Junge, Jürgen Volz, and Arno Rauschenbeutel. "Fiber-optical switch controlled by a single atom." Physical review letters 111, no. 19 (2013): 193601. DOI: 10.1103/PhysRevLett.111.193601

[15] Wu X, Tong L. Optical microfibers and nanofibers. Nanophotonics. 2013 Dec 16;2(5–6):407-428. https://doi.org/10.1515/nanoph-2013-0033

[16] Monifi, Faraz, Sahin Kaya Özdemir, Jacob Friedlein, and Lan Yang. "Encapsulation of a fiber taper coupled microtoroid resonator in a polymer matrix." IEEE Photonics Technology Letters 25, no. 15 (2013): 1458-1461. DOI: 10.1109/LPT.2013.2266573

[17] Buck JA. Fundamentals of optical fibers. John Wiley & Sons; 2004 Apr 27.

[18] Okamoto K. Fundamentals of optical waveguides. Academic press; 2006.

[19] Love JD, Henry WM, Stewart WJ, Black RJ, Lacroix S, Gonthier F. Tapered single-mode fibres and devices. Part 1: Adiabaticity criteria. IEE Proceedings J (Optoelectronics). 1991 Oct 1;138(5): 343-354. DOI: 10.1049/ip-j.1991.0060

[20] Black, R. J., S. Lacroix, F. Gonthier, and J. D. Love. "Tapered single-mode fibres and devices. II. Experimental and theoretical quantification." IEE Proceedings J-Optoelectronics 138, no. 5 (1991): 355-364. DOI: 10.1049/ip-j.1991.0061

[21] Sumetsky, M. "Nanophotonics of optical fibers." Nanophotonics 2, no. 5-6 (2013): 393-406. DOI: 10.1515/nanoph-2013-0041

[22] Kakarantzas, George, Sergio G. Leon-Saval, T. A. Birks, and P. St J. Russell. "Low-loss deposition of solgel-derived silica films on tapered fibers." Optics letters 29, no. 7 (2004): 694-696. DOI: 10.1364/OL.29.000694

[23] Li, Bao-li, Jin-hui Chen, Fei Xu, and Yan-qing Lu. "Periodic micro-structures in optical microfibers induced by Plateau-Rayleigh instability and its applications." Optics express 25, no. 4 (2017): 4326-4334. DOI: 10.1364/OE.25.004326

[24] Vélez-Cordero, J. Rodrigo, A. M. Velázquez-Benítez, and J. Hernández-Cordero. "Thermocapillary flow in glass tubes coated with photoresponsive layers." Langmuir 30, no. 18 (2014): 5326-5336. DOI: 10.1021/la404221p

[25] Xu, Z. Y., Y. H. Li, and L. J. Wang. "Versatile technique to functionalize optical microfibers via a modified sol-gel dip-coating method." Optics letters 39, no. 1 (2014): 34-36. DOI: 10.1364/OL.39.000034

[26] Velázquez-Benítez, Amado M., Moisés Reyes-Medrano, J. Rodrigo Vélez-Cordero, and Juan Hernández-Cordero. "Controlled deposition of polymer coatings on cylindrical photonic devices." Journal of Lightwave Technology 33, no. 1 (2014): 176-182. DOI: 10.1109/JLT.2014.2377173

[27] Landau, Levich, and B. Levich. "Dragging of a liquid by a moving plate." In Dynamics of curved fronts, pp. 141-153. Academic Press, 1988. DOI: 10.1016/B978-0-08-092523-3.50016-2

[28] De Ryck, Alain, and David Quéré. "Inertial coating of a fibre." Journal of Fluid Mechanics 311 (1996): 219-237. DOI: 10.1017/S0022112096002571

[29] Quéré, David. "Fluid coating on a fiber." Annual Review of Fluid Mechanics 31, no. 1 (1999): 347-384. DOI: 10.1146/annurev.fluid.31.1.347

[30] White, David A., and John A. Tallmadge. "A theory of withdrawal of cylinders from liquid baths." AIChE Journal 12, no. 2 (1966): 333-339. DOI: 10.1002/aic.690120223

[31] Rayleigh, Lord. "On the instability of jets." Proceedings of the London mathematical society 1, no. 1 (1878): 4-13. DOI: 10.1112/plms/s1-10.1.4

[32] Li, Bei-Bei, Qing-Yan Wang, Yun-Feng Xiao, Xue-Feng Jiang, Yan Li, Lixin Xiao, and Qihuang Gong. "On chip, high-sensitivity thermal sensor based on high-Q polydimethylsiloxane-coated microresonator." Applied Physics Letters 96, no. 25 (2010): 251109. DOI: 10.1063/1.3457444

[33] Yang, Min K., Roger H. French, and Edward W. Tokarsky. "Optical properties of Teflon® AF amorphous fluoropolymers." Journal of Micro/Nanolithography, MEMS, and MOEMS 7, no. 3 (2008): 033010. DOI: 10.1117/1.2965541

[34] Hoang, Van Thuy, Grzegorz Stępniewski, Karolina H. Czarnecka, Rafał Kasztelanic, Van Cao Long, Khoa Dinh Xuan, Liyang Shao, Mateusz Śmietana, and Ryszard Buczyński. "Optical properties of buffers and cell culture media for optofluidic and sensing applications." Applied Sciences 9, no. 6 (2019): 1145. DOI: 10.3390/app9061145

[35] Moutzouris, Konstantinos, Myrtia Papamichael, Sokratis C. Betsis, Ilias Stavrakas, George Hloupis, and Dimos Triantis. "Refractive, dispersive and thermo-optic properties of twelve organic solvents in the visible and near-infrared." Applied Physics B 116, no. 3 (2014): 617-622. DOI: 10.1007/s00340-013-5744-3

Chapter 3

Optical Inhouse Networks

Ulrich H.P. Fischer, Matthias Haupt and Peter Kußmann

Abstract

Optical fiber networks are currently the standard for delivering high bandwidth to customers. Various access technologies to business networks with a very high bandwidth up to access networks for buildings and individual consumers have emerged. In the area of business networks, bandwidths of 10 Gb/s have become established, while in the area of customer bandwidths of 100 Mb/s to 1 Gb/s are used. This chapter will focus on the optical network connections inside buildings. The use of optical glass fibers or/and polymeric optical fibers in different network topologies in connection to high-speed actual WIFI- technologies will be discussed.

Keywords: optical fiber, distributed network structure, optical polymeric fibers, local area networks, open building reference model

1. Introduction

Depending on how far the glass fiber extends into the access network, one speaks of "Fiber to the Curb" (FTTC), "Fiber to the Building" (FTTB), "Fiber to the Home" (FTTH), "Fiber to the Desk" (FTTD). As an alternative, DSL technology, outdoor DSLAMs and VDSL with vectoring, contribute data transmission rates of up to 200 Mbit/s for broadband distribution to the subscriber, whereby these copper-based connection technologies have reached now their capacity limits [1].

As shown in **Figure 1**, the international optical network is essentially divided into three levels. Based on the international level of the global area network with

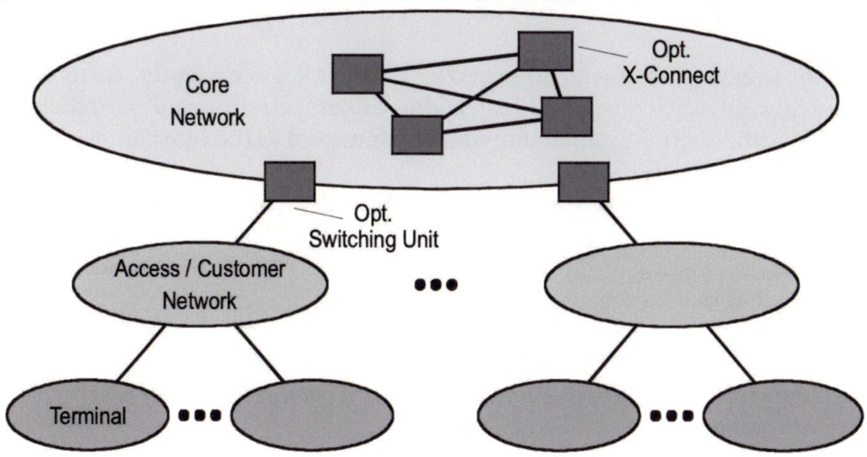

Figure 1.
Worldwide optical network structure.

very high data rates of 10–100 Gb/s, branches go to the level of regional networks with data rates of 1–10 Gb/s and from there to the third level of the customer network. There we speak of data rates of 100 Mb/s to 10 Gb/s. Based on this core network, the wireless networks (4G, 5G) to supply the mobile devices in the area with up to 1Gb/s are realized [2].

The mobile networks with LTE or 5G and the cable television networks also use this term for the part of their networks that includes the subscriber connections and offers access to higher network levels. However, radio technology can only be seen as an alternative to fiber optic infrastructure in rural areas, as little cabling needs to be installed. Unfortunately, the bandwidth of these radio networks decreases quadratic by the distance to the radio base station, so that the effectively transferable data rate will fall down strictly in rural areas. The cable television networks are currently based on DOCSYS 3.1 and already use fiber optic connections as FTTB variants and are therefore to be seen as fiber optic technology despite different software and hardware configurations. Currently, most of the optical fiber optic connections are made in the FTTB area [3, 4].

At a connection point in the basement of the building, the fiber optic cable coming from the outside is connected to a router or switch. Starting from there, the data is transferred to the different areas and floors of the office building or the apartment block either via fiber optics, polymer fibers or electrical Ethernet cables (e.g. CAT 7). This is referred as network levels 3 and 4, before branching from the different floors to the individual offices or apartments via switches. The other cabling in the apartments and offices is called network level 5. Here, in turn, there are appropriate routers for the apartments or switches for the offices in order to distribute the data in the individual rooms. Levels 3–5 are nowadays mainly connected via optical cables and will be analyzed in this chapter in a differentiated manner and discussed with examples. This also includes the use of additional technologies such as WiFi, Zigbee, Bluetooth or dLAN/powerline [5] in the home and office [6–8].

2. Distributed network structures

Computer networks in apartments, small buildings and office environments (Small Office/Home Office - SoHo) are typical fields of application for Ethernet-based communication networks [9]. Local services such as network printers or file shares are implemented on these network structures, or DSL and cable providers offer their customers IP-based services such as Internet access, IP-TV or VoIP (Voice over IP).

For the implementation of Ethernet/POF-based SoHo networks, different areas must be conceptually considered (e.g. laying cables, selection and structuring of electronic components), implemented and connected to the Internet using suitable devices and processes [10–13].

Based on the sub-areas to be implemented for the realization of a network in the desired SoHo target market, the protocols and standards to be considered are so diverse that the existing orientations as well as the derived test methods and devices appear unsuitable.

For the design of a general and expandable structure in the SoHo environment, an abstraction model was developed - which is suitable as a basis for orientation for the implementation of network structures, service models and test methods in the SoHo environment.

In the telecommunication networks a special scheme is applied to distribute the data from the source/headend to the customer. The distribution is divided into five levels (see **Table 1** and **Figure 2**):

Network level	Task
1	Content production
2	Operation of the head-end stations that receive and forward the TV and radio signals
3	Street/curb distributors
4	House distributors
5	Between router in the home and telephone/Internet devices

Table 1.
Network levels in the telecommunication network.

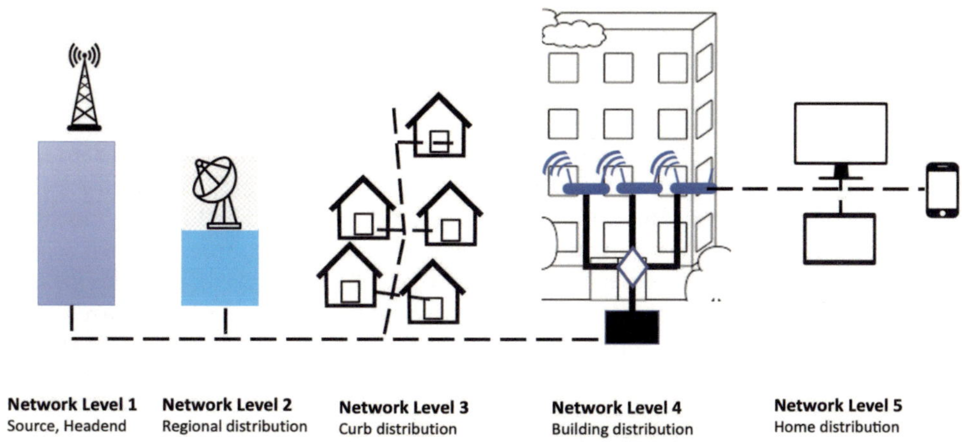

Figure 2.
Distribution of the telecommunications transmission path from the provider via the curb distribution to the building and in the building to the flats.

Level 1: Content/data production.

These include program providers such as Premiere, the public and private broadcasting stations and also Cloud applications.

Level 2: Operation of head-end stations.

The internet data is received via these stations and then passed on. In the case of the TV cable network, one of these head-end stations is always required, as this is where the signals from the satellite are converted and processed for cable reception, among other things. The reception technology in the cable networks has been completely converted to the digital standard since DOCSYS 3.1 [14] and is therefore at the same level as the telecommunications networks. Typical data rates are between 10 Gb/s up to 100Gb/s.

Level 3: Street distributor/curb.

This refers to the network areas that were relocated from the head-end stations to the residential areas (FTTC-fiber to the curb). Typical data rates are between 1 Gb/s up to 10Gb/s.

Level 4: house distributor.

Customers can only be reached via the fourth network level into the houses (FTTB - fiber to the building). Many small operators are exclusively active here, the number of which is estimated at several thousand. In order to offer new products and services, the operator must be able to feed them into the network. This requires an adaptation of several network levels, which, however, often turns out to be very difficult due to the large number of responsible companies at the various levels. Only in the rarest of cases does an operator have several levels, which makes the actual structure difficult to understand for the end user. Typical data rates are between 100 Mb/s up to 10Gb/s.

Level 5: Flat distribution.

The network connection directly with the customer is now being rolled out worldwide via optical fibers, since only these technologies enable a correspondingly high bandwidth. This level is between router and telephone/Internet devices such as smartphones, tablets or IP TV realized. Typical data rates are between 10 Mb/s up to 1Gb/s (FTTH - fiber to the home).

3. Open building interconnection reference model - OBI

Based on the different orientations of the four identified work areas (building blocks) (see **Figure 3**) as well as the multitude of standards and guidelines to be taken into account, which data center operators are familiar with - but not known to house and apartment owners in the targeted SoHo target market - the following situation arises:

A model is required that includes the four identified work areas - defines interfaces and thus offers orientation for further work [15, 16]. All areas relevant for the conception, construction and operation of SoHo networks and in-house communication can be structured, edited and tested in a reproducible manner [17, 18].

Based on the basic services defined within the OBI model, the conception of test scenarios was started. For this it is necessary to classify the network structure that is likely to be encountered. Starting from a transfer point (e.g. DSL [19], FTTH, DOCSIS [20] that connects to the public wide area data network (WAN) [18], SoHo routers are used that provide services for the internal network that are defined using an operating system (firmware) (LAN) (e.g. switch, WLAN, FXS (Foreign eXchange Station), FITH (Fiber In The Home)). The owner of the SoHo network has no influence on the implementation of the WAN area, apart from the choice of provider and product. The situation is different with the implementation of the LAN structure. Which implementation is used primarily defines the intended use of the LAN. The router already mentioned shows the following structure (**Figure 4**) taking into account its functions (**Figure 5**):

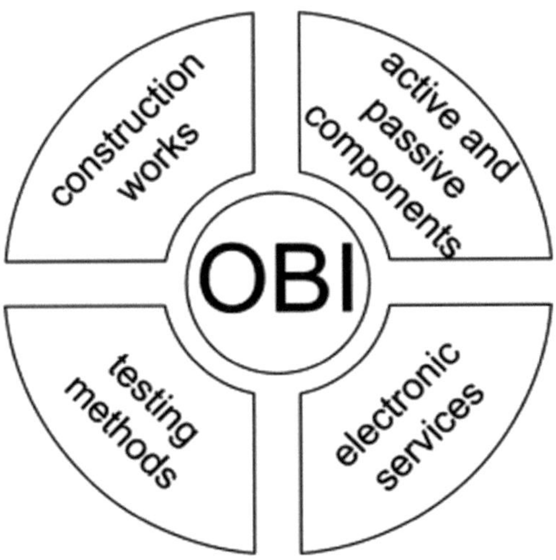

Figure 3.
O B I - open building reference model.

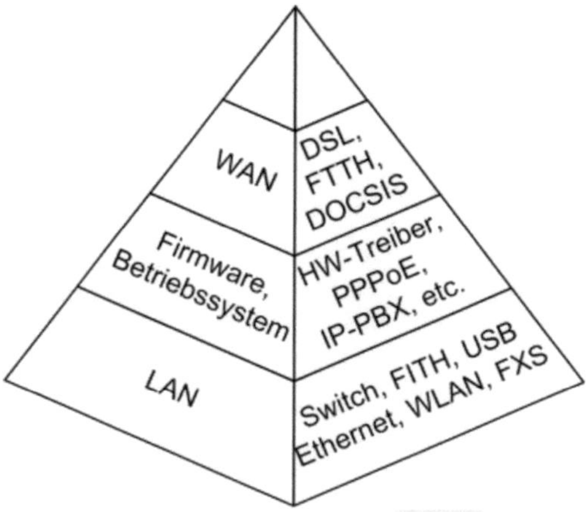

Figure 4.
Functional overview of a SoHo router.

	WAN	Router	LAN
Symbol			
Funktion	DSL, FTTH, DOCSIS	HW-Treiber, PPPoE, IP-PBX, Firewall	Switch, FITH, Ethernet, WLAN, FXS

Figure 5.
Network structure in the SoHo-environment.

Due to the owner's preference, all technical options (e.g., WLAN, fiber optic, POF, Cat [5–7], etc.) as well as the associated active and passive components are available for a structure.

If the SoHo network is viewed from the point of view of the services operated on it, the guarantee of the correct functioning and the quality of service by the provider of the WAN connection ends in the SoHo router which is depicted in **Figure 6**. It is assumed here that the provider of the WAN connection terminates the VoIP traffic on a SIP registrar [21] integrated in the SoHo router. All local end devices contact this SIP registrar, which in turn forwards the VoIP traffic to the telephone server of the WAN provider for termination using a SIP trunk.

The configurations, implementations and functions in the LAN area for mapping the functional correctness and quality of service for the defined basic services (network access, VoIP, WLAN) are part of the current work. For the investigation of the bandwidth-prioritized VoIP, a test scenario specially adapted to SoHo environments was designed.

In order to obtain reproducible results, the dependency on external disturbances (on the WAN side) must be excluded, which is why a local VoIP registrar was configured (codecs: g.711u [22], h264 [23]) and used. The SIP video telephones use - for the VoIP within the LAN - different transmission media, e.g., Cat (5/6/7), POF and WLAN and different combinations and configurations of the active and passive components. The ITU standards for PESQ [24] and the E-model [25] are used for objective assessment of the voice quality of VoIP calls.

Both test methods take into account all parameters involved in the transmission (e.g., noise, SNR, latencies, jitter, echoes, packet losses, etc.) as well as their mutual dependency. For the examination, a defined language file is transmitted (see **Figure 7**- red/

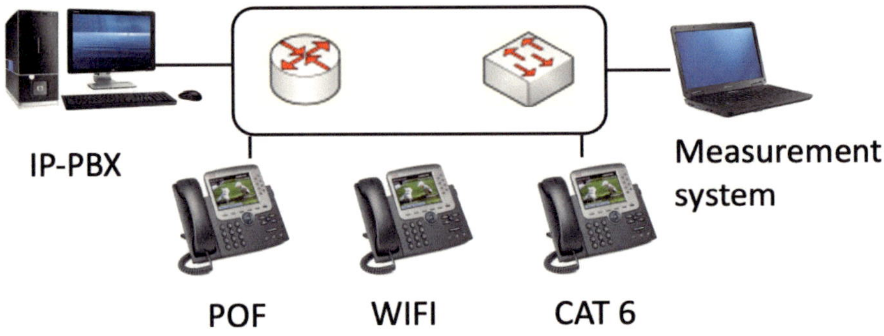

Figure 6.
SoHo - network topology for VoIP test.

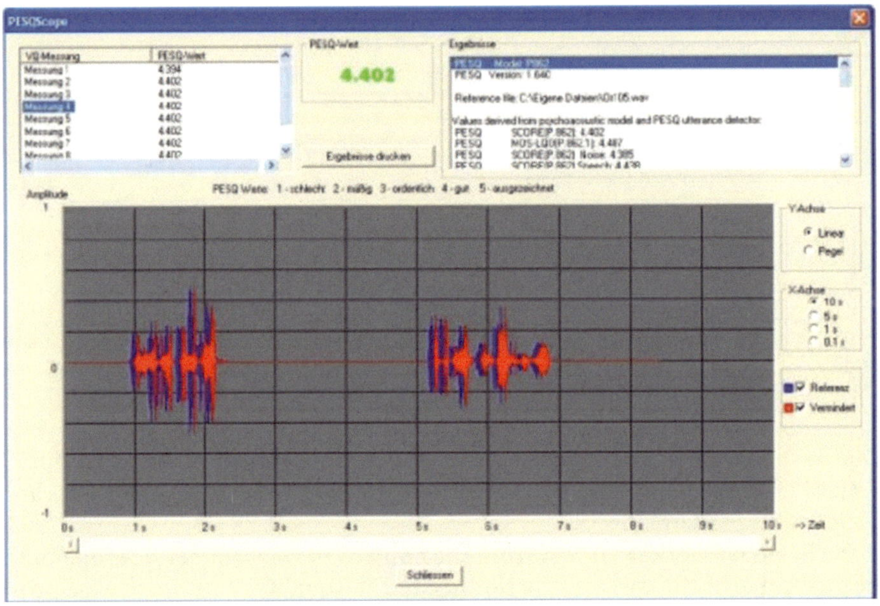

Figure 7.
PESQ-measurement in SoHo network topology.

reduced) and compared with the original file (see **Figure** 7- blue/reference). The result is a numerical score value (PESQ-MOS factor for PESQ, R factor for E-model) for the comparison tables.

After calibration of the individual transmission links and selection of suitable test parameters and routines, the configurable parameters of the active components (e.g., QoS, VLAN, IPv6) can be examined in the context of "Influence on the quality of VoIP transmissions in SoHo environments". Statements can also be made about the influence of the codecs used (e.g., H264, G711, G722 [26]) or the number of maximum SIP connections in the context of the bandwidth limit of the existing network topology. By using the VoIP registrar function in commercially available SoHo routers that are already available on the market (today already with all major Internet providers), the test routines can be run through - without measurement setups and configurations - and thus provide comparable results. By using the OBI model, communication and network structures in SoHo environments, especially for technology-supported care assistance systems [27] with the personal data that arise there, can be referenced and verified.

4. Network types in the building

4.1 Glass fibers vs. polymeric fibers vs. CAT

Optical waveguides are made of optical glasses or assembled, partially assembled, provided with plug connections cables and lines for the transmission of light in the visible as well as ultraviolet or infrared range. Fiber optic cables form flexible connections for the transmission of optical signals. Depending on the application, the fiber optic cables consist of Quartz glass, e.g. pure silicon dioxide (SiO2) or organic glass Polymeric fibers consist of acrylic glass [28]. From a physical point of view, both optical waveguides are dielectric waveguides.

Today, fiber optic cables [29] are mainly used as a transmission medium for wired telecommunication processes. In addition, there are diverse applications: fiber optic cables for laser radiation for material processing [30, 31], in medicine for lighting and imaging purposes: microscope lighting [32], endoscopes [33], decoration lighting [34], for contact-free sensors [35], in measurement technology, e.g. in infrared thermometers and spectrometers [36].

Today, fiber optic cables are increasingly used for information transmission, in telecommunications and also in the area of computer networking. The term optical fiber is standardized in DIN 47002 and VDE 0888 and means that it is a conductor in which modulated light is transmitted. The fiber optic cable can be made of fiberglass or plastic. With plastic fiber-optic cables, the so-called POF, high transmission rates can be achieved, which can be up to several million bit/s. Furthermore, POF are insensitive to electromagnetic interference, largely secure against eavesdropping and have very low attenuation values compared to copper conductors.

A comparison of different fiber optical waveguides in glass and in polymeric materials are depicted in **Figure 8**. There are multimode fibers (MMF) available in two sizes, 62.5 or 50 microns, and four classifications: OM1 (62.5/125 μm), OM2, OM3, OM4 (50/125 μm) [37]. The GOF fiber type for SoHo applications is the multimode GOF with a diameter 50 μm core and 125 μm cladding. The bandwidth of this device is typically 1–10 GBit/s over 100–500 m. The POF has a very limited bandwidth of 1 Gbit/s over 100 m link length in Ethernet networks.

As depicted in **Figure 9** the single mode GOF (SMF) offers the highest bandwidths of over 10 Gbit/s, which go well beyond the bandwidths required in the

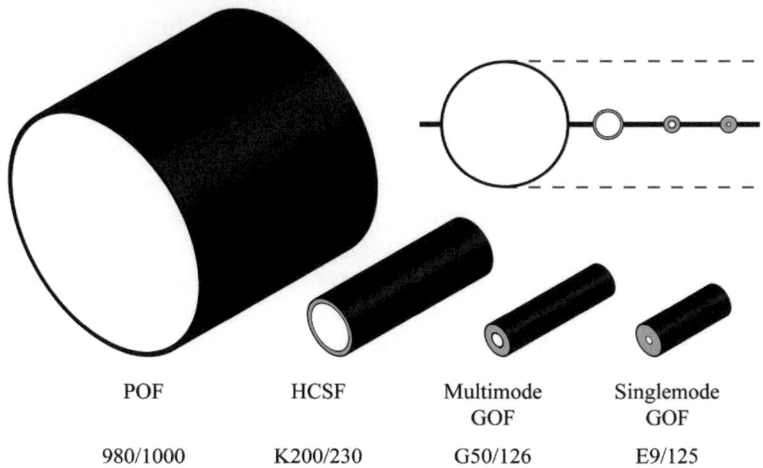

Figure 8.
Dimensions of GOF and POF fiber types.

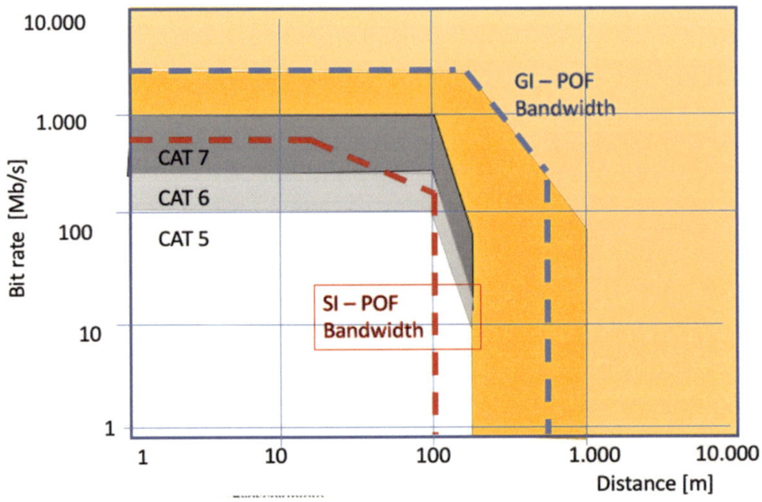

Figure 9.
Transmission speeds of single mode and multimode GOF, POF and CAT for inhouse applications.

SoHo area. This is why the SMF still plays a subordinate role in the short-term computing area in buildings. On the other hand, the multimode GOF (MMF) with bandwidths of 1 Gbit/s has significantly higher application potential, since the connectors and other active components such as transmitters and receivers are significantly cheaper than those of SMF components. The POF has significantly expanded its bandwidth potential in recent years and with gradient index POF (GI-POF) can also allow up to 10 Gbit/s over 50 m transmission distance, but has only been investigated in research studies to date. The step index POF (SI-PF), which can be used commercially with a maximum of 1 Gbit/s, is significantly worse in terms of transmission bandwidth. On the other hand, experimental studies show, that Wavelength Division Multiplex (WDM) techniques [38] applied in the POF spectrum of 400 nm to 780 nm can overcome the bandwidth restrictions and can realize more than 15 Gb/s via 4 chromatic transmission channels [39]. Additional WDM sources can extend the overall bandwidth to more than 40 Gb/s [40].

The copper technology of the CAT cables currently also achieves gigabit transmission speeds, but is very susceptible to installation errors such as bending radii that are too small. However, the CAT cable connection has the advantage of passive networking without further active transmitter/receiver elements.

4.2 POF fibers for inhouse applications

For signal transmission over short to medium distances of up to approx. 100 m, optical waveguides made of acrylic glass (polymethyl methacrylate or PMMA), so-called POF, are used.

Polymer fiber technology for optical data transmission has developed very dynamically over the past 10 years [41]. Starting with simple transmission options for the consumer sector such as digital links between DVD players and preamplifiers in the home multimedia sector (TOSLINK [42] system) with data rates of a few Mb/s, the technology has now established itself in the automotive sector with the use of MOST bus [43]. Here, POF is used in the visible wavelength range, since the components at this level of application must be manufactured as cost-effectively as possible for the end user. Mobile multimedia applications are of particular importance in the automotive sector, where over 50 vehicle types (approx. 15 million vehicles) have been equipped with POF bus systems since its introduction in 2001.

In addition to the higher data rate and the resulting improved integration of multimedia applications in busses or automobiles, considerable weight reductions in the cable of 30% are also achieved [44, 45].

For these reasons, optical data transmission is increasingly being used in close proximity, e.g. in office and house communication, in production facilities, in medical technology or in bus systems for cars, trains and planes.

In the following, the optical basics of fibers, called POF (polymer optical fibers) for short, their active and passive components for network technology and their fields of application in the in-house area are presented.

4.3 Optical properties and advantages of POF

Optical fibers consist of a highly transparent core, a cladding and a protective coating and/or buffer. The light-guiding core is used to transmit the signal. The cladding has a lower optical refractive index (density) than the core. As a result, the cladding causes total reflection at the boundary layer and thus guiding the radiation in the core of the optical waveguide. However, light can also get into the cladding through bending or coupling at the beginning of the route. This is usually undesirable and the jacket and protective coating are therefore designed in such a way that this light is strongly attenuated.

The outer protective coating helps against mechanical damage and protects the fiber from environmental influences. The POF consists of PMMA (acrylic glass), has a core diameter of approx. 1 mm (**Figures 10** and **11**) and has a bandwidth of

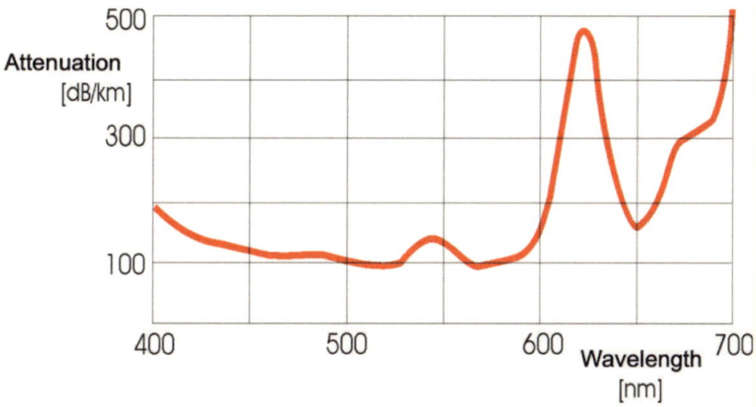

Figure 10.
Attenuation diagram of POF in the visible regime.

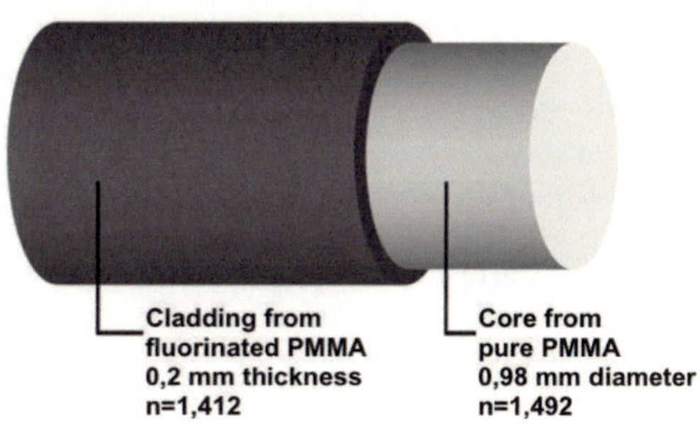

Figure 11.
Schematic draw of a polymeric fib.

	POF	GOF	CAT
EMV	++	++	—
Galvanic isolation	++	++	—
Bug security	+	+	—
Risk in explosive environments	++	++	+
Low weight	++	+	—
Small bending radius	+	—	—
Flexibility	+	—	+
Low cost	++	—	+
Bandwidth	+	++	+
Attenuation/m	—	++	—
Cable and connector assembly costs	++	—	+
++ very good, + good, – unsatisfactory.			

Table 2.
Comparison of GF, CAT and POF transmission connections in SoHo environment [46].

100 Mb/s over 100 m, which can be expanded to 1000 Mb/s with special modulation techniques and laser transmitters. The advantages of POF in laying technology are obvious in several areas compared to WiFi, powerline/dLAN or CAT solutions (see **Table 2**, [46]):

1. Practical aspects

    - Very inexpensive, easy to work with
    - Low weight, very small diameter: 1/10 of copper cables
    - More flexible and cheaper than glass fiber optic cables
    - Easy handling
    - EMC insensitivity

2. Security aspects

    - Better insurability
    - Security against eavesdropping
    - Short-circuit protection, free of hum loops

3. Technical aspects

    - Data rates can be expanded in a future-proof manner (investment security)
    - Several signals can be transmitted on one fiber
    - Significantly cheaper and easier to lay than glass fibers

The 1 mm fiber type is the cheapest to manufacture and is therefore used in 95% of all commercial applications. The refractive index is constant over the entire core cross-section (SI-POF). With other fiber types, for example gradient index fibers (GI) or multi-level index fibers (MSI), significantly higher bandwidths (currently 2.3GHz/100m) can be achieved in the laboratory, but these do not play a role in the consumer market segment. The advantages of the standard SI-POF lie in the wide availability, the very low price, the favorable attenuation behavior (**Figure 10**) a very high numerical aperture, which enables a simple and effective coupling and extraction of light.

5. Usable connectors for glass fibers and polymeric fibers in the home

The optical polymer fiber POF has particularly good and simple properties for the connection technology between fibers and optical transmitters and receivers [47]. Due to its simple structure, which with a core of 0.98 mm and a cladding of 0.2 mm corresponds exactly to 1 mm outer diameter, the cable can be cut straight with a cutter knife (see illustration) very easily. In contrast to fiber optic connections, the separation and cutting of a plastic fiber is much easier because the fiber optics require very complex mechanics and the dismantling of the fiber optic cables (see YouTube video [48]). A connection of two fibers has particularly low losses, since the large diameter reduces the mechanical boundary conditions for the accuracy of the alignment to + −0.1 mm. This corresponds to a 100 times lower necessary accuracy compared to a fiber optic connector. Because the alignment accuracy between the fiber and the optical element can be measured very generously without generating large losses, the plug-in connections are correspondingly easy to assemble. The **Figure 12** shows an overview of the typical plug connections for POF cables.

A distinction is made between metallic and plastic plugs. The metallic plugs are used in the area of rough environments like car entertainment busses or factory environments for stable connections. The plastic plugs were used in the area of in-house networks without large load peaks.

No plugs are typically used for use in the home, while the fibers can be simply cut off (see **Figure 14**) and butt-inserted into the receptacle of a media converter. This saves a lot of costs and unnecessary time for the connector assembly, as well

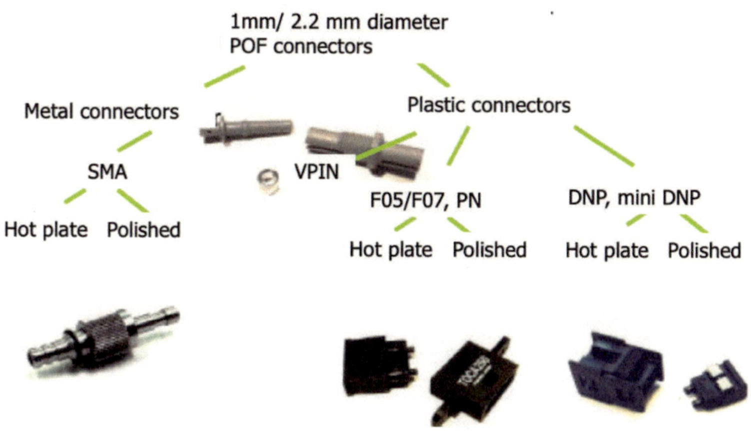

Figure 12.
Typical plugs for POF connections.

as a simple possibility for the non-specialist to lay network cables himself without special knowledge. Typical connectors for butt-inserted connections are shown in **Figure 15**.

Another area for the introduction of POF network technology is the construction and connection technology, e.g. of splicing for POF. This would make it possible to directly couple POF without additional components. This technology is established in fiber optic technology, but in a technology that cannot be used for POF. That is why it would be of crucial importance for the cable laying technology of POF to have such a connection technology with innovative approaches. A splice kit (**Figure 13**) has been implemented at the HarzOptics GmbH Company with industrial partners using injection molding technology, which provides very good conditions for an easy-to-use splice [49]. Very low attenuation of 0.2 dB is typical for this spice method. The basic idea is that the fibers are fit very exact into the splice core of 0.98 mm and the core hoe was filled before with glue. The refractive index of the glue is 1,5 and the hardening of the glue is realized by the use of a UV light lamp, which took only 3 minutes to be hardened (**Figure 15**).

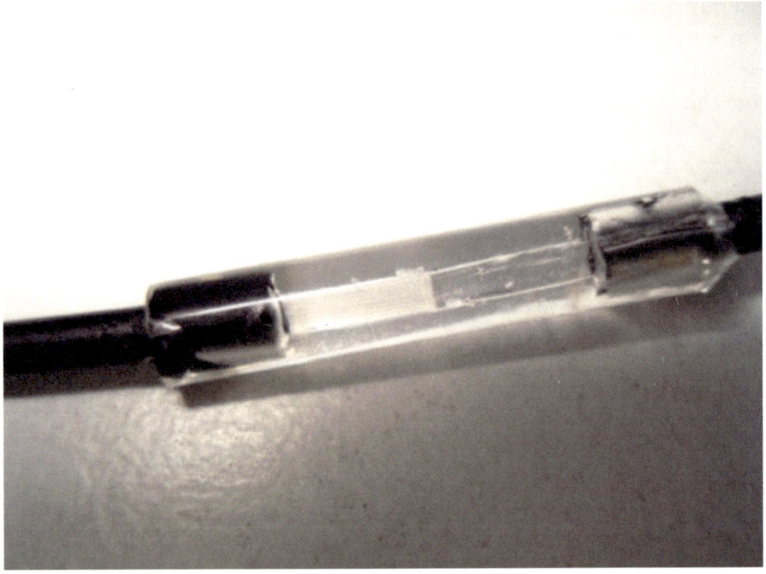

Figure 13.
Optical splice with low attenuation.

Figure 14.
POF fiber cutting easy with cutter knife.

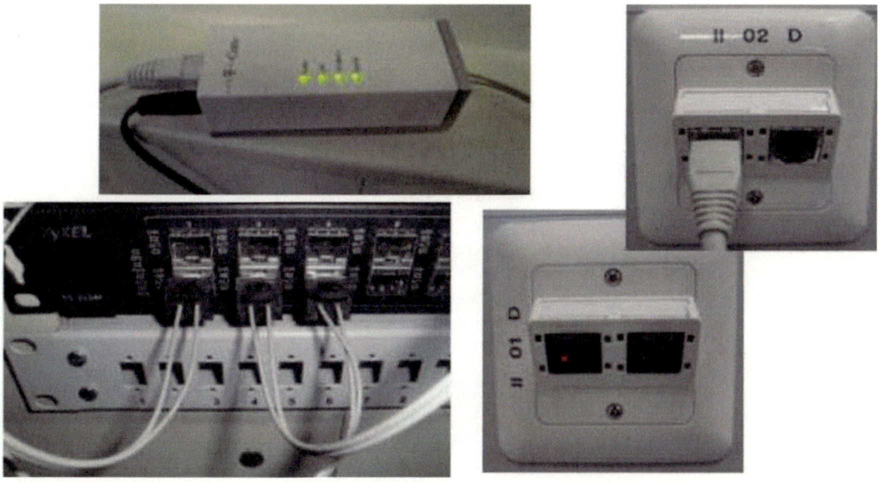

Figure 15.
POF media convertors, Ethernet switches and 4-port optical POF switch RJ-45 wall outlets with WiFi-Accesspoint (Rutenbeck GmbH).

For the home sector, also known as domotics, an increasing demand for bandwidth is to be expected over the next few years. One reason for this development is the triple play promoted by the leading telecommunications companies, which means a bundled range of services such as IP telephony, IP TV and the Internet (**Figure 16**). Another term used in this context is the "active house". The development of this concept represents the integration of communication and entertainment, as well as the active control of all functional processes in the house (control of the heating, blinds, monitoring systems, etc.), also known as building automation.

For these areas, active components on the one hand, e.g., media converters for setting up dynamic network structures, and on the other hand, passive optical components such as splitters for the inexpensive construction of such networks are available in sufficient numbers for Ethernet applications up to 1000 Mb/s. There are a variety of applications and simple installation techniques for POF media converters or POF adapters (**Figure 15**) and POF Ethernet switches for installation in flush-mounted switch boxes are available for setting up home networks with polymer fibers [50]. On the user side, one or more ports with RJ-45 interfaces (10/100/1000Base-Tx) are available for connecting the end devices. The polymer fiber is connected on the installation side. The POF is connected to the optical interfaces (1000Base-Fx) using plug-in terminals.

The switch shown also allows the construction of star, bus, tree and ring structures with polymer fiber cables. Some of the ports of POF Ethernet switches even offer Power over Ethernet (PoE) functionality, so that IP telephones, IP cameras or WLAN access points with IEEE 802.11n data rates (up to 240 Mb/s net data rate [51]) can be operated on the POF network without plug-in power supplies.

6. Example of an in-house network with optical POF fibers

A typical example of an application in a single-family home is shown in **Figure 16**. The data which is supplied from outside from the so-called network level 3 via fiber optics or DSL to the house is sent to a router at the house transfer point.

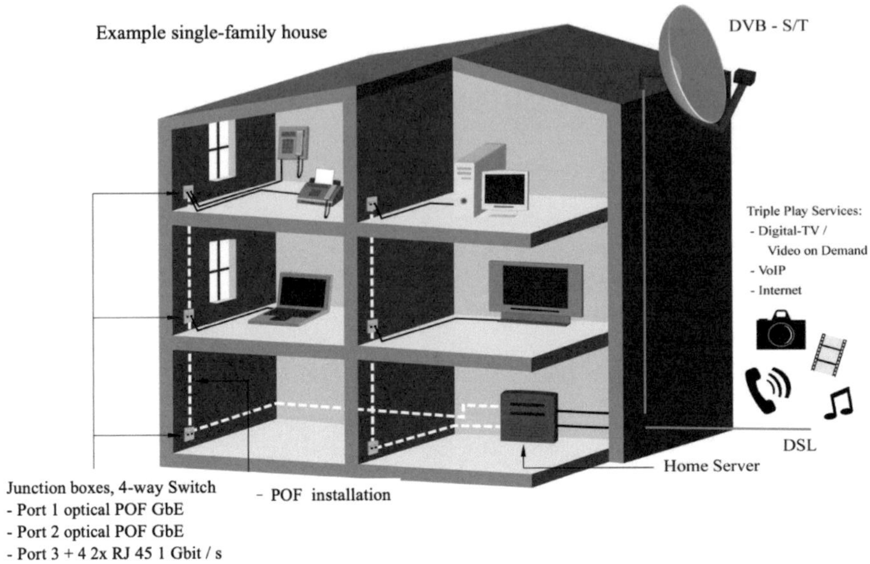

Figure 16.
Optical POF bus in the home network.

From this router, optical POF fiber cables are installed in all rooms in the house with active switches with four ports and connected there. Forwarding via additional switches the data can easily be routed to each room and connected. Furthermore, a corresponding WiFi module can be integrated in these switches, which can illuminate each room individually with small radio energy and is available for mobile devices in this room. By reducing the radio energy, each room can be connected to its own WiFi radio network and thus does not interfere with the transmission quality of the adjoining rooms. The radio energy must be set so low that the radio waves from the individual WiFi areas in the room do not get into the neighboring room and are attenuated enough by the walls.

It is also possible to lay the polymer fiber slightly behind baseboards in existing structures and thus not have to finance complex and expensive construction measures.

7. Conclusions

Optical fiber networks are currently the standard for delivering high bandwidth to customers. It was discussed, that there are various access technologies to local networks with a very high bandwidth up to access individual customers in their homes or flats. The use of optical glass fibers or/and polymeric optical fibers in different network topologies in connection to high speed actual WiFi- technologies have been discussed.

Both, the copper networks with CAT connections, as well as the networking with optical glass fiber and optical were compared and their strengths and weaknesses were shown. The Polymer Optical Fiber exhibits many advantages in comparison to glass fiber and copper as the medium for communication. The mentioned applications show different special sectors to the application of one of the three transmission technologies.

The focus in this work lies on the possibility of conveying high data rates, as well as the simplest possible relocation of network components in the SoHo area. In the area of permanently installed network components, the use of POF fibers proved to be particularly suitable for network bandwidths of up to 1 Gb/s. In addition to the almost relocated optical components, the installation of a wireless network via WiFi is a particularly good addition to make mobile devices easily networkable for the customer. Thus, a recommendation can be issued for both, optical polymer fibers with Ethernet network technology in combination with current WiFi technology. Both system components will experience further expansion stages in the range in the next few years and thus always remain applicable and expandable.

Author details

Ulrich H.P. Fischer[1*], Matthias Haupt[2] and Peter Kußmann[1]

1 Harz University of Applied Sciences, Wernigerode, Germany

2 Jade University of Applied Sciences, Wilhelmshaven, Germany

*Address all correspondence to: ufischer@hs-harz.de

References

[1] European Commission. Broadband: Technology overview [Internet]. Broadband: Technology overview. 2020. p. 2. Available from: https://digital-strategy.ec.europa.eu/en/policies/broadband-technology-overview

[2] Shafi M, Molisch AF, Smith PJ, Haustein T, Zhu P, De Silva P, et al. 5G: A Tutorial Overview of Standards, Trials, Challenges, Deployment, and Practice. IEEE J Sel Areas Commun. 2017;35(6):1201-21.

[3] German Telecom. Fiber to the X - The architecture of fiber-optic lines [Internet]. 14.06.2021. 2021. p. 3. Available from: https://www.telekom.com/en/company/details/ftth-fttb-fttc-fiber-optic-lines-in-fiber-optic-networks-608068

[4] Grötschel M, Raack C, Werner A. Towards optimizing the deployment of optical access networks. EURO J Comput Optim [Internet]. 2014;2(1):17-53. Available from: https://doi.org/10.1007/s13675-013-0016-x

[5] Sagar N. Powerline communications systems. 2011. 147 p.

[6] Suryo Prayogo S, Al Rafi F, Mukhlis Y. Design and Built IoT Home Panic Button for Smart City. In: Journal of Physics: Conference Series. 2019.

[7] Mendes TDP, Godina R, Rodrigues E, Matias J, Catalão J. Smart Home Communication Technologies and Applications: Wireless Protocol Assessment for Home Area Network Resources. Energies. 2015;8:7279-311.

[8] Danbatta SJ, Varol A. Comparison of Zigbee, Z-Wave, Wi-Fi, and Bluetooth Wireless Technologies Used in Home Automation. In: 2019 7th International Symposium on Digital Forensics and Security (ISDFS). 2019. p. 1-5.

[9] DIN EN 50173: Information technology - Generic cabling systems,. 2010 p. 20.

[10] Dave Hood ET. Introduction. In: Gigabit-Capable Passive Optical Networks [Internet]. John Wiley & Sons, Ltd; 2012. p. 1-8. Available from: https://onlinelibrary.wiley.com/doi/abs/10.1002/9781118156070.ch1

[11] Dave Hood ET. System Requirements. In: Gigabit-Capable Passive Optical Networks [Internet]. John Wiley & Sons, Ltd; 2012. p. 9-47. Available from: https://onlinelibrary.wiley.com/doi/abs/10.1002/9781118156070.ch2

[12] Dave Hood ET. Optical Layer. In: Gigabit-Capable Passive Optical Networks [Internet]. John Wiley & Sons, Ltd; 2012. p. 49-125. Available from: https://onlinelibrary.wiley.com/doi/abs/10.1002/9781118156070.ch3

[13] Dave Hood ET. Transmission Convergence Layer. In: Gigabit-Capable Passive Optical Networks [Internet]. John Wiley & Sons, Ltd; 2012. p. 127-218. Available from: https://onlinelibrary.wiley.com/doi/abs/10.1002/9781118156070.ch4

[14] Fischer UHP, Schulze H-M, Reinboth C. NGA im ländlichen Raum: Erfahrungsbericht aus der Harzregion NGA in rural areas: Experiences from the Harz region. In: ITG-Fb 262: Breitbandversorgung in Deutschland Beiträge der 10 ITG-Fachkonferenz, 18-19 April 2016 in Berlin [Internet]. Berlin: VDE Verlag , Berlin; 2016. p. 5. Available from: https://www.vde-verlag.de/buecher/454193/itg-fb-262-breitbandversorgung-in-deutschland.html

[15] Zygiaris S. Smart City Reference Model: Assisting Planners to

Conceptualize the Building of Smart City Innovation Ecosystems. J Knowl Econ [Internet]. 2013;4(2):217-31. Available from: https://doi.org/10.1007/s13132-012-0089-4

[16] Borgia E. The Internet of Things vision: Key features, applications and open issues. Comput Commun [Internet]. 2014;54:1-31. Available from: https://www.sciencedirect.com/science/article/pii/S0140366414003168

[17] Tucker WD, Blake EH, Marsden G. Open User Interconnect and Quality of Communication. In: D. Browne (ed.), editor. Southern African Telecommunication Networks and Applications Conference (SATNAC). Stellenbosch, South Africa; 2004. p. 261-262.

[18] Gazis V, Görtz M, Huber M, Leonardi A, Mathioudakis K, Wiesmaier A, et al. A survey of technologies for the internet of things. In: 2015 International Wireless Communications and Mobile Computing Conference (IWCMC). 2015. p. 1090-5.

[19] ITU G.992.1: Asymmetric digital subscriber line (ADSL) transceivers.

[20] ITU J.122: Second-generation transmission systems for interactive cable television services - IP cable modems Recommendation.

[21] Jabel AN, Manickam S, Ramdas S. A study of SIP trunk security and challenges. In: 2012 IEEE International Conference on Electronics Design, Systems and Applications (ICEDSA). 2012. p. 239-43.

[22] ITU-T G711. ITU G.711: Pulse code modulation (PCM) of voice frequencies [Internet]. ITU-G Recommendations; 2012 p. 10. Available from: https://www.itu.int/rec/T-REC-G.711

[23] ITU-T Recommendations. ITU H264 standard [Internet]. ITU-T Recommendations; 2012 p. 10. Available from: https://www.itu.int/ITU-T/recommendations/rec.aspx?rec=11466

[24] ITU P.862: Perceptual evaluation of speech quality (PESQ): An objective method for end-to-end speech quality assessment of narrow-band telephone net-works and speech codecs.

[25] ITU G.107: The E-model: a computational model for use in transmission planning,.

[26] ITU. ITU G.722: 7 kHz audio-coding within 64 kbit/s [Internet]. ITU-G Recommendations; 2012 p. 10. Available from: https://www.itu.int/rec/T-REC-G.722/_page.print

[27] Fischer-Hirchert UHP, Hoppstock S, Kußmann P. Intelligent Assistance Systems with Real-Time Sensor Data Analysis-Framework for Digital Care. German Society for Biomedical Engineering, editor. Biomed Eng-Biomed Tech [Internet]. 2020;65((s1)):2. Available from: https://www.edas.info/p27318

[28] CME edu. Acrylic glas properties [Internet]. 2020. Available from: https://www.cs.cmu.edu/afs/cs/academic/class/15294t-s18/resources/Acrylic-data-sheet.pdf

[29] Imani R, Cuellar GH. Introductory Chapter: Optical Fibers. In: Huerta-Cuellar G, Imani R, editors. Optical Fiber Applications [Internet]. Rijeka: IntechOpen; 2020. Available from: https://doi.org/10.5772/intechopen.91397

[30] Fischer UHP, Krips O, Muller E, Jacob a, Müller E, Jacob a. Laser microwelding for fiber-chip coupling modules with tapered standard monomode fiber ends for optical communication systems. Opt Eng

[Internet]. 2002;41(12):3221-9. Available from: http://www.scopus.com/inward/record.url?eid=2-s2.0-0036980315&partnerID=40&md5=2eb8dc59f97d555bdaf5af63609ebc95

[31] Katayama S. Fundamentals and Details of Laser Welding. Berlin: Springer Berlin Heidelberg; 2020. p 198.

[32] Gu M, Sheppard CJR, Gan X. Image formation in a fiber-optical confocal scanning microscope. J Opt Soc Am A [Internet]. 1991 Nov;8(11):1755-61. Available from: http://josaa.osa.org/abstract.cfm?URI=josaa-8-11-1755

[33] Smithwick QYJ, Seibel EJ. Depth enhancement using a scanning fiber optical endoscope. In: Alfano RR, editor. Optical Biopsy IV [Internet]. SPIE; 2002. p. 222-33. Available from: https://doi.org/10.1117/12.465249

[34] Křemenáková D, Militky J, Meryova B, Lédl V. Characterization of Side Emitting Polymeric Optical Fibres. J Fiber Bioeng Informatics. 2012;5:423-31.

[35] Abang A, Webb DJ, Peng G-D. Strain response of POF sensors. In: 22nd International Conference on Optical Fiber Sensors. 2012.

[36] Höll S, Haupt M, Fischer UHP. Design and Development of a injection-molded demultiplexer for optical communication systems in the visible range. Appl Opt [Internet]. 2013;52(18):4103-10. Available from: http://www.ncbi.nlm.nih.gov/pubmed/23842150

[37] G.651 I-T. Characteristics of a 50/125 µm multimode graded index optical fibre cable. 1993.

[38] Haupt M, Fischer UHPHP. WDM over POF - the inexpensive way to break through the limitation of bandwidth of standard POF communication. Proc SPIE Photonics Packag Integr Interconnects VII [Internet]. 2007;6478(6478):64780I-64780I – 10. Available from: http://proceedings.spiedigitallibrary.org/proceeding.aspx?articleid=1298065

[39] Joncic M, Kruglov R, Haupt M, Caspary R, Vinogradov J, Fischer UHP, et al. Four-Channel WDM Transmission Over 50-m SI-POF at 14.77 Gb/s Using DMT Modulation. Photonics Technol Lett IEEE2. 2014;26(13):1328-31.

[40] Fischer UHP, Haupt, Joncic. WDM over POF up to 40 Gbit/s. In: Fraunhofer-Gesellschaft. Dr. Tolga Tekin, editor. 1st Optical Interconnect in Data Centers Symposium [Internet]. Berlin, Germany 2014,: Fraunhofer-Gesellschaft. Dr. Tolga Tekin,; 2014. p. 6. Available from: http://www.phoxtrot.eu/1st-optical-interconnect-in-data-centers-symposium/

[41] Fischer-Hirchert UHP. Beyond the Dispersion Limit of Standard Polymeric Fiber Transmission Systems. In: Prof Abraham Katzir, editor. OASIS 7 (Optical Engineering and Science in Israel) Proceedings [Internet]. Telaviv; 2019. Available from: http://oasis7.org.il/wp-content/uploads/2019/03/Oasis-7-Poster-List-3.pdf

[42] Toshiba Corp. Fiber-Optic Devices TOSLINKTM [Internet]. 2008 p. 16. Available from: http://www.digikey.com/Web Export/Supplier Content/Toshiba_264/PDF/Toshiba_BCE0037_catalog.pdf?redirected=1

[43] Cooperation TM. MOST Technology Report [Internet]. Karlsruhe; 2016. Available from: www.mostcooperation.com

[44] Fischer UHP, [Fischer11] U. H. P. Fischer-Hirchert, M. Haupt, and M. Joncic, Fischer UHP. Optical Transmission Systems Using Polymeric Fibers. In: Predeep P, editor. Optoelectronics- Devices and Applcations [Internet]. Rijeka, Croatia:

intech open; 2011. p. Chapter 22. Available from: http://www.intechopen.com/articles/show/title/optical-transmission-systems-using-polymeric-fibers

[45] Grzemba. MOST. The Automotive Multimedia Network. From MOST25 to MOST150. Poing: Franzis; 2011. 250 p.

[46] Visani D. Fiber-Optic Technologies for Wireline and Wireless In-building Networks [Internet]. alma; 2012. Available from: http://amsdottorato.unibo.it/4535/

[47] Poisel H, Ziemann O. What are POF ? [Internet]. 2015. Available from: https://www.th-nuernberg.de/fileadmin/institute/pof-ac/pof-ac_docs/Was_ist_POF_2017_A5__2017.pdf

[48] Fischer UHP. Dismantling and cutting glass fiber [Internet]. 2010. Available from: https://youtu.be/NZVH8yJcjCM

[49] Fischer UHP, Haupt M. Optische Schlüsselemente für die Polymerfaserübertragung mit Mikrospritzguss. In: Mageburger Maschinenbautage 2009. Magdeburg: Otto von Guericke Universität; 2009. p. 10.

[50] homefibre. POF Gigabit Switches [Internet]. 14.06.2021. 2021. Available from: https://www.home-fibre.ch/shop/en/pof-switches/96-12-4-port-gigabit-pof-switch.html

[51] IEEE. IEEE 802.11n [Internet]. 2009. Available from: https://standards.ieee.org/standard/802_11n-2009.html

Section 2

Optical Fiber in Communications

Chapter 4

Coded Modulation and Impairment Compensation Techniques in Optical Fiber Communication

Zhipei Li, Dong Guo and Ran Gao

Abstract

This chapter deals with coded modulation and impairment compensation techniques in optical fiber communication. Probabilistic shaping is a new coded modulation technology, which can reduce transmission power by precoding, reduce bit error rate and improve communication rate. We proposed a probabilistic shaping 16QAM modulation scheme based on trellis coded modulation. Experimental results show that this scheme can achieve better optical SNR gain and BER performance. On the other hand, in order to meet the demand of transmission rate of next generation high speed optical communication systems, multi-dimensional modulation and coherent detection are sufficiently applied. The imperfect characteristics of optoelectronic devices and fiber link bring serious impairments to the high baudrate and high order modulation format signal, causes of performance impairment are analyzed, pre-compensation and receiver side's DSP techniques designed for coherent systems are introduced.

Keywords: Coherent Optical Communication, Coded Modulation, Probabilistic Shaping, Digital Signal Processing, Pre-Compensation, Quadrature Amplitude Modulation

1. Introduction

With the rapid development of Internet services, higher requirements are put forward for the transmission rate, system capacity and stability of communication. Optical fiber communication has become one of the main communication methods in the world because of its large transmission bandwidth, long-distance transmission and strong anti-interference ability.

The optical fiber communication systems are mainly divided into two types: intensity modulation-direct detection (IM-DD) and coherent optical communication systems. The IM-DD system is mainly used in access networks, passive optical networks (PON) and data centers, and its transmission distance is usually less than 80 km. Specified in the standard, the highest transmission rate with single-wavelength of IM-DD system is 100Gb/s, transmitting 56GBaud 4-level pulse amplitude modulation (PAM4) signal. The coherent optical transmission system, which uses multi-dimensional modulation to improve spectrum efficiency, local oscillator lasers to increase sensitivity, and digital signal processing (DSP) technology for impairment compensation, thereby can greatly improve transmission performance, increase transmission rate and distance, and is usually used in in large-capacity long-distance backbone networks and metropolitan area networks.

Quadrature amplitude modulation (QAM) is commonly used in coherent optical fiber transmission system. In this method, evenly distributed constellation points are arranged in two-dimensional space to form a constellation diagram. The performance loss of the modulation scheme tends to $\pi e/6$ (1.53 dB) from the Shannon limit. To further approach Shannon limit capacity and improve system performance, probabilistic shaping technology can be used [1]. Probabilistic shaping is a new coded modulation technology. It can reduce transmission power by precoding, reduce bit error rate and improve communication rate without changing the original system. And with the change of channel environment, the channel can be matched by changing the size of the shaping, thus improving the flexibility of the system.

The future research direction of high-speed optical fiber communication is digital coherent optical communication technology [2–4]. With the development of high-bandwidth optoelectronic devices, digital-to-analog converter (DAC), analog-to-digital converter (ADC) and application specific integrated circuit (ASIC) chips, beyond 800Gb/s transmission with single wavelength above becomes possible. In order to achieve such high-speed transmission, DSP technology plays an important role in dealing with chromatic dispersion (CD), polarization mode dispersion (PMD), frequency offset and phase noise, by compensating the signal at the transmitter and receiver in the electrical domain. With highly integrated and flexible digital coherent optical detection technology, high-speed, large-capacity and long-distance optical communication can be effectively realized.

The most important technologies in coherent optical communication system are introduced in this chapter, a novel coded modulation technology based on probabilistic shaping and DSP-based Impairment compensation techniques.

2. Novel coded modulation technology based on probabilistic shaping

2.1 Principle of probabilistic shaping

The main idea of probabilistic shaping is to reduce the probability of occurrence of constellation points in outer ring, increase ones in the inner ring, and change the constellation of uniform distribution into non-uniform distribution. The common distribution matcher is constant composition distribution matcher (CCDM), which makes the occurrence probability of each constellation point conform to Maxwell Boltzmann distribution:

$$P_X(x_i) = \frac{e^{-v|x_i|^2}}{\sum_{j=1}^{m} e^{-v|x_j|^2}} \tag{1}$$

Where, $X = \{x_1, x_2, \ldots \ldots, x_m\}$ is the constellation symbol set, v is the probability distribution factor, and the value range is $0 \sim 1$. The larger value of v, the higher the degree of constellation shaping [5].

Taking 16QAM modulation format as an example, the constellation probability distribution diagram is shown in **Figure 1**, and the constellation diagram after recovery at the receiving end is shown in **Figure 2**.

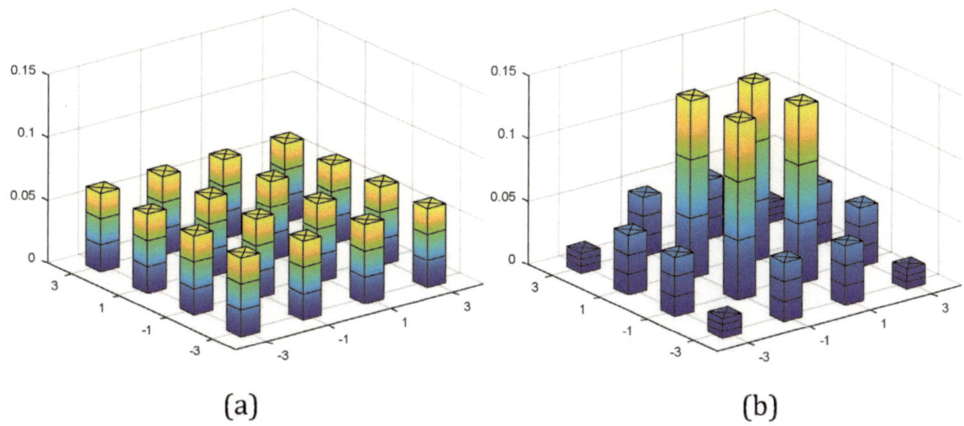

Figure 1.
Probability distribution of 16QAM signal. (a) Uniform 16QAM; (b) PS 16QAM.

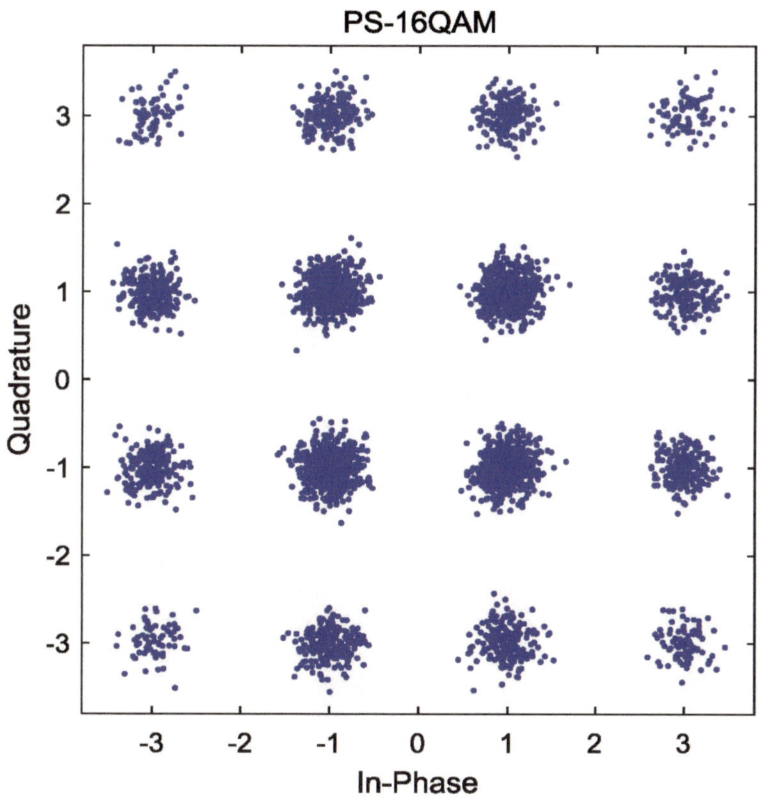

Figure 2.
Constellation for receiver recovery using probabilistic shaping technique.

2.2 Experiment

To evaluate the performance of the proposed scheme, an experiment is carried out by employing the coherent optical communication system setup illustrated in **Figure 3**. The system parameters are shown in **Table 1**. At the transmitter, a 1550 nm lightwave with power of 10 dBm and line-width of 100 kHz is employed as the laser source, followed by a polarization beam splitter to divide the output light

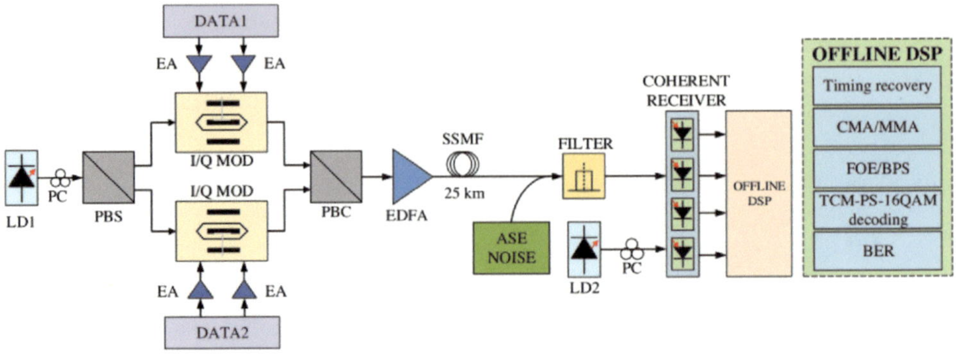

Figure 3.
Experimental setup for PDM single carrier system.

Parameter	Specification
Center wavelength	1550 nm
Laser linewidth	100 kHz
Input power of fiber	0 dBm
Amplification	EDFA
EDFA gain	20 dB
EDFA noise figure	4 dB
Fiber	SSMF
Attenuation	0.2 dB/km
Nonlinear coefficient	$1.3(\text{W}\cdot\text{km})^{-1}$
Chromatic dispersion	17 ps/(nm·km)
Span length	25 km

Table 1.
System parameters.

into orthogonal polarized pair. Then two I/Q modulator are applied to modulate two orthogonal light waves, respectively. Trellis-Coded Modulation (TCM)-PS-16QAM signals are generated offline by MATLAB program. And each modulator is driven by two 10Gbaud amplified electrical TCM-PS-16QAM signals with a frame length of 33,336 symbols, and 10 patterns (333,360 symbols) are collected for bit error ratio (BER) calculation. The peak-peak voltage of the two amplified electrical signals are both set as 2.0 V. And a polarized beam combiner is applied to combine two orthogonal modulated lightwaves, which is amplified by an erbium-doped fiber amplifier (EDFA) in the standard single mode fiber (SSMF) for transmission. The gain and the noise figure of the EDFA are 20 dB and 4 dB, respectively. The input power of the fiber is 0 dBm, and the fiber is a SSMF with attenuation of α = 0.2 dB/km, nonlinear coefficient of $\gamma = 1.3\ (\text{W}\cdot\text{km})^{-1}$, and dispersion of D = 17 ps/nm/km. At the receiver, the ASE noise is added to the received optical signal to adjust the received optical signal noise ratio (OSNR) with the resolution of 0.1 nm. The optical signal added with noise is bandpass-filtered and converted into an electrical signal by a coherent receiver. The laser diode (LD) generates a local oscillating light with the power of 5 dBm and linewidth of 100 kHz. In the offline DSP module, timing recovery is implemented by Gardner algorithm and chromatic dispersion is compensated digitally. To process different shaped signals, a pre-convergence constant modulus algorithm with step of 2e-6 and taps of 9 followed by 1% pilot aided multi-modulus algorithm are used for polarization demultiplexing and compensation of polarization mode dispersion without training sequence. In addition, fast Fourier transform based frequency offset estimation (FFT-FOE) algorithm and blind phase search algorithm are applied to realize frequency offset estimation and compensate the laser phase noise, respectively. Finally, TCM-PS-16QAM decoding and statistics of BER are implemented.

To verify our proposed scheme, we compare the performance of 8QAM, TCM-16QAM-4state, TCM-16QAM-8state, TCM-16QAM-16state, and TCM-PS-16QAM-4state respectively. The convolutional encoders and information entropy parameters are displayed in **Table 2**. Here, some modulation formats have different entropy, and lower entropy means a higher baud-rate, which makes their bit-rates consistent.

In the experiment, the TCM-PS-16QAM (H = 2.8 bits/symbol) with OSNR = 15 dB is firstly transmitted. After the optical transmission system, the received signal of the subset S0, S1, S2, S3 and the overall constellation at the receiver is shown in **Figure 4**. The successful transmission of the TCM-PS-16QAM signal means that our proposed novel scheme is reasonable and realizable.

Scheme	Entropy (bits/symbol)	Baud rate (GBaud)
8QAM	3	10
TCM-16QAM-4state	3	10
TCM-16QAM-8state	3	10
TCM-16QAM-16state	3	10
TCM- PS-16QAM-4state	2.9	10.4
	2.8	10.7
	2.7	11.1

Table 2.
Modulation format parameters.

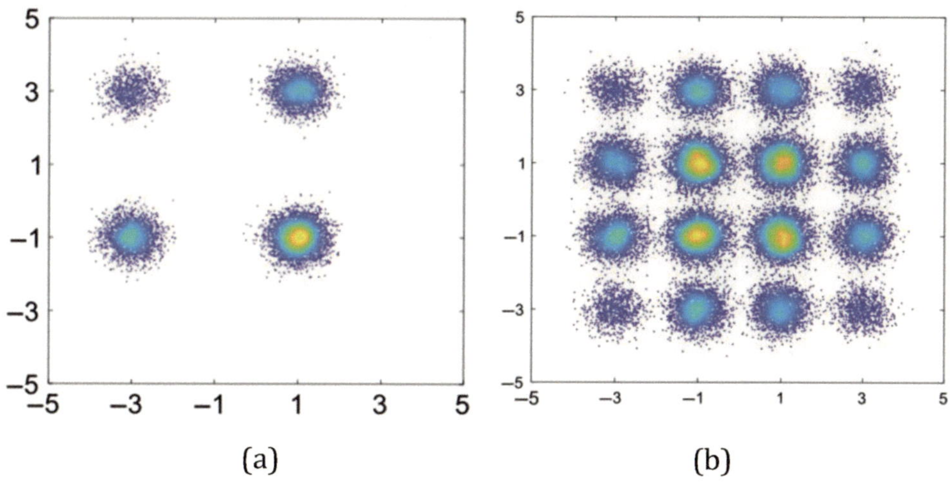

Figure 4.
Probability distribution with the entropy of 2.8 bits/symbol of the proposed (a) subset S_0 constellation, (b) TCM-PS-16QAM constellation at the receiver.

Meantime, it can be seen that the probability distribution of the constellation points achieves the goal of low energy point density and high energy point sparseness.

Figure 5 depicts the BERs of 8QAM, TCM-16QAM-4state, TCM-16QAM-8state, TCM-16QAM-16state, and TCM-PS-16QAM-4state (H = 2.9 bits/symbol) with the same data rate of 60 Gb/s after 25-km SSMF transmission. It can be seen that when the OSNR is above 5 dB, the performance of TCM-16QAM-nstate (n = 4, 8, 16) is better than 8QAM. And with the increase of OSNR, the BERs of TCM-16QAM-nstate (n = 4, 8, 16) fall faster than 8QAM. At the BER of 1×10^{-3}, TCM-16QAM-4state obtains the gain of 3.4 dB compared to 8QAM. However, TCM-16QAM-8state slightly outperforms TCM-16QAM-4state and the OSNR gain is just 0.4 dB. Meantime, compared to 8 state, the gain of TCM-16QAM-16state grows only 0.4 dB too. And the decoding complexity double as the number of state double. So, increasing the number of states is not a suitable way to obtain coding gain, especially when the coding gain is close to the limit. The required OSNR for BER of 1×10^{-3} is about 7.7 dB for TCM-16QAM-4state, and 6.8 dB for TCM-PS-16QAM-4state (H = 2.9 bits/symbol), The OSNR improvement is increased by 0.9 dB with a little more complexity. Compared to TCM-16QAM-8/16state, TCM-PS-16QAM-4state (H = 2.9 bits/symbol) has lower decoding complexity and better performance. And the gain of TCM-PS-16QAM-4state (H = 2.9 bits/symbol) grows 0.5 dB and 0.1 dB, respectively. This is mainly due to the shaping gain brought by PS. Under the condition that the minimum Euclidean distance in the constellation

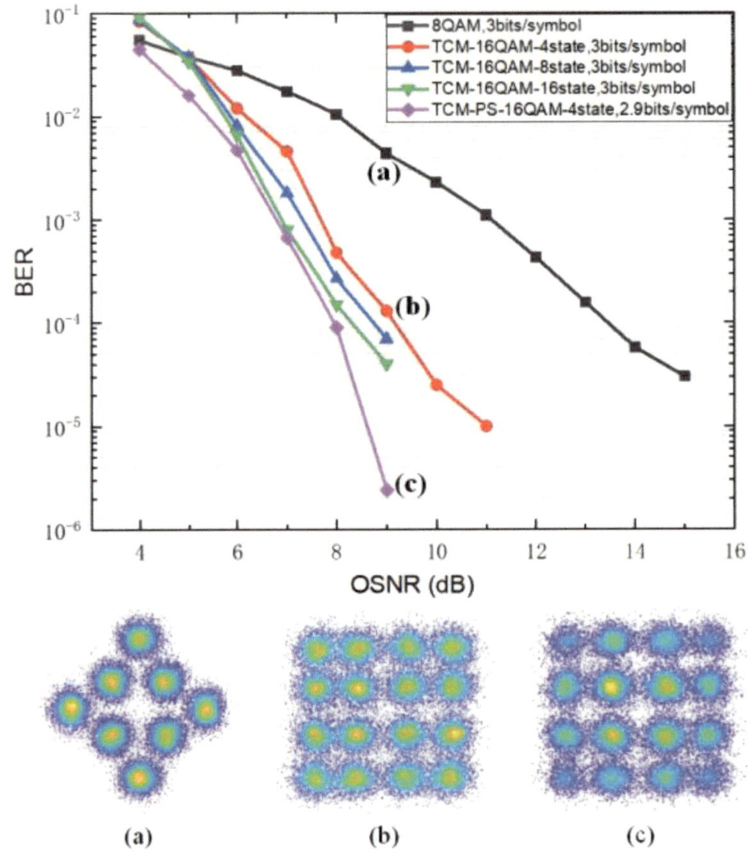

Figure 5.
BER curves and constellations of different modulation formats (H = 2.9 bits/symbol) for 25-km transmission. (a)8QAM, (b)TCM-16QAM-4/8/16state, (c)TCM-PS-16QAM-4state.

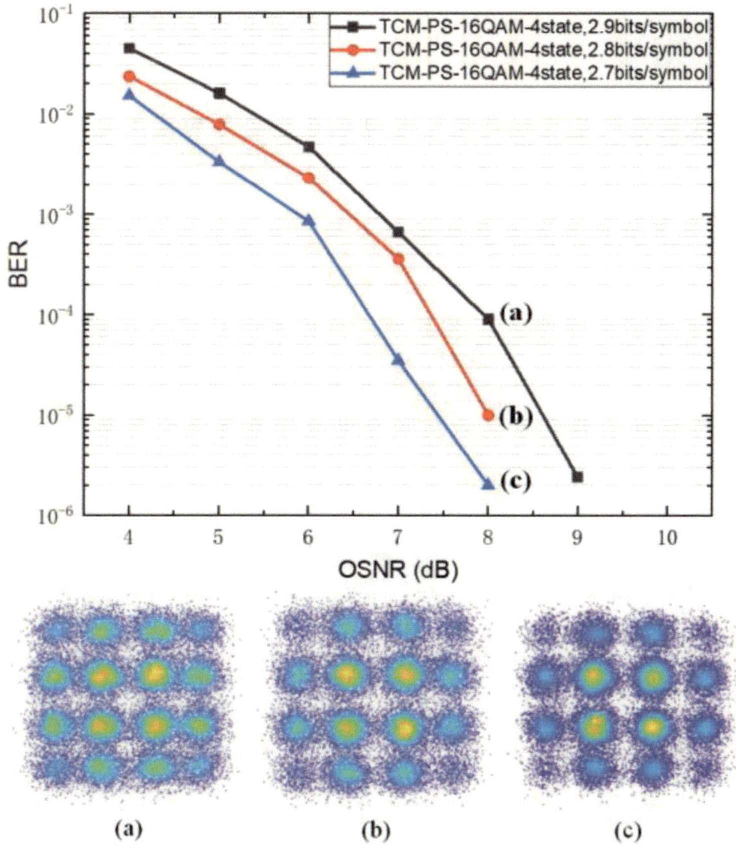

Figure 6.
BER curves and constellations of TCM-PS-16QAM-4state for 25-km transmission. (a) H=2.9 bits/symbol, (b) H = 2.8 bits/symbol, (c) H = 2.7 bits/symbol.

remains unchanged, the convergence of most constellation points greatly reduces the average power of the constellation, improving the performance of the entire modulation system.

The BER curves of TCM-PS-16QAM-4state with different information entropy after 25-km transmission are also measured, and the measured results are shown in **Figure 6**. In our experiments, according to their different information entropy, the baud rate is set differently, their information entropy is 2.9, 2.8, 2.7 bits/symbol, so the baud rate is 20.7, 21.4, 22.2Gbaud, namely, the bit rate is 60 Gb/s. It can be seen that the required OSNRs are 6.8 dB, 6.5 dB, and 5.9 dB at the BER of 1×10^{-3}. In other words, TCM-PS-16QAM-4state (H = 2.7 bits/symbol) obtains OSNR gain of 0.6 and 0.9 dB compared with TCM-PS-16QAM-4state (H = 2.8 and 2.9bits/symbol), respectively. This advantage proves that as the information entropy decreases, more shaping gains can be obtained. Obviously, at the same OSNR, the lower the information entropy, the better the BER performance and the more flexible information entropy and gain compared to the traditional TCM-16QAM.

3. Impairment compensation techniques in coherent communication

Coherent optical communication systems are implemented in the form of optical modules in commercial communication equipment [6]. As shown in **Figure 7**, The

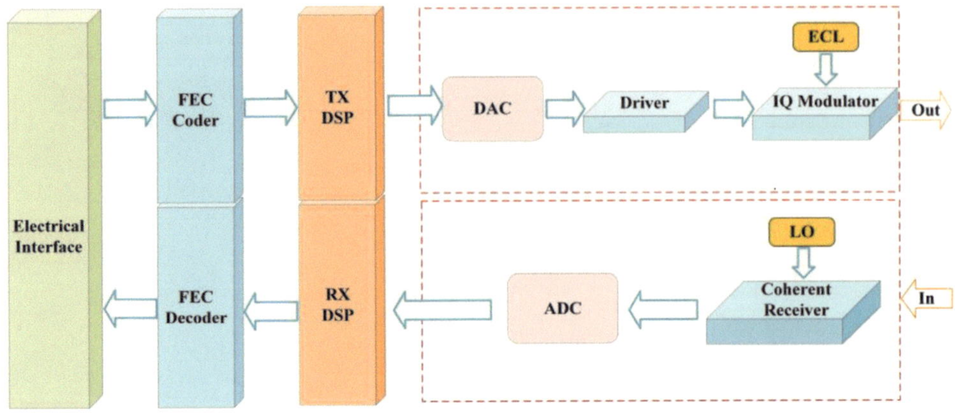

Figure 7.
Coherent optical module structure.

transmitted information is encoded by forward error correction (FEC) through the electrical interface and then transmitted to transmitter-side (TX) DSP. The TX DSP includes constellation mapping, pulse shaping, pre-equalization, skew compensation, dispersion pre-compensation, etc. The main purpose of implementing DSP-based pre-compensation techniques at the transmitter is to increase the signal-to-noise ratio (SNR) of the transmitted signal while the transmission power is limited, reduce the bandwidth requirement for signal transmission, avoid inter-symbol interference (ISI), and reduce the impairment caused by the optoelectronic devices.

After single mode fiber channel, the coherent optical signal is transmitted into the coherent receiver and ADC to obtain four electrical signals, which consist of dual polarizations X/Y and in-phase/quadrature (IQ) components XI, XQ, YI, and YQ. The receiver-side (RX) DSP contains a series of impairment compensation and equalization algorithms for the optoelectronic devices and fiber link, including IQ signal orthogonalization, normalization, dispersion compensation, clock recovery, polarization demultiplexing, frequency offset estimation, and phase noise recovery. DSP compensation technology at the receiver side is the core of coherent optical communication. It can recover multi-dimensional modulated signals from distorted constellation, to realized large-capacity transmission.

This chapter will introduce the impairment compensation technology based on DSP in two parts, all common compensation techniques are described in detail.

3.1 Impairment compensation techniques at transmitter side

Pre-equalization and skew compensation are basic DSP techniques at transmitter side, pre-equalization technique is used to compensate the filtering effect caused by the bandwidth limitation of transmitting devices, and skew compensation is to compensate the delay of XI, XQ, YI and YQ signals while passing through the electrical and optical paths [7]. At the same time, in order to cope with more and more high-speed transmission, such as 400Gb/s, 800Gb/s and 1.2 Tb/s transmission, high-order modulation format and high baud rate signal is generated and transmitted, such as 96Gbaud 32QAM and 80Gbaud 64QAM, which is extremely sensitive to linear and nonlinear impairment of optoelectronic devices, therefore, other pre-compensation algorithms such as look-up table (LUT) and digital predistortion (DPD) are usually added to the next generation coherent optical module's DSP algorithm. In this section, basic principles of the impairment compensation techniques at transmitter side are described for the development of next generation coherent optical transmission systems.

3.1.1 Bandwidth limitation and pre-equalization

Restricted by material and technical level, the frequency response of the opto-electronic device is not flat in the range of the bandwidth needed to transmit the signal, as shown in **Figure 8**, the transmitted signal will fade at high frequency, which leads to inter symbol interference, the smaller the bandwidth is, the more serious the ISI is and the worse the BER performance.

To suppress the performance degradation of high-speed signal caused by bandwidth limited system, the pre-equalization technology based on DSP can alleviate the bandwidth shortage of the transmitter device, which is an effective bandwidth compensation method. To implement pre-equalization, it is necessary to obtain the frequency response of the transmitter, including DAC, electric driver, modulator and so on. First, we send specific training sequence X without any compensation, then receive the signal Y with a high bandwidth digital sampling oscilloscope, we can estimate the frequency response H of the transmitter by comparing the transmitted and received signals with least square (LS) algorithm

$$H_{LS} = \left(X^H X\right)^{-1} X^H Y \tag{2}$$

Then we multiply the inverse of the estimated frequency response with the transmitted data in the frequency domain

$$X_{TX} = X \cdot H_{LS}^{-1} \tag{3}$$

Thus, the high frequency component of the signal can be raised at the transmitter to resist the low-pass filtering effect of the device. The spectrum of the transmitted signal is flat to reduce the inter symbol crosstalk.

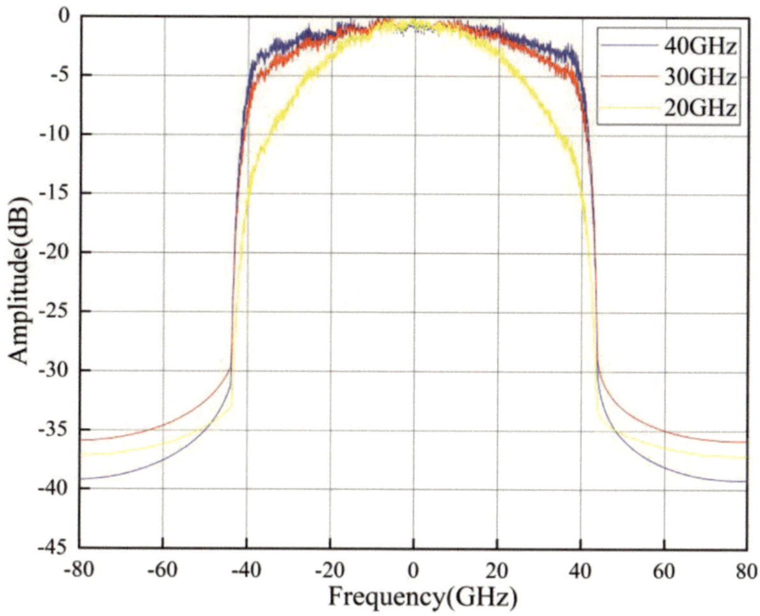

Figure 8.
Spectrum of transmitted signal with different bandwidth limitation.

3.1.2 Look-up table algorithm

LUT is a pre-compensation method, which is used to compensate the memory effect of the amplifier. From the time domain point of view, the memory effect means that the current output symbol of the amplifier is not only dependent on the current input symbol, but also related to the past input symbol value. From the perspective of frequency domain, memory effect can be defined as the phenomenon that the amplitude and phase characteristics of intermodulation distortion term of amplifier change with the variation of envelope frequency of input signal. Considering the influence of $2N+1$ symbols before and after the symbols at the intermediate time, all possible transmitted sequences are $X(k-N:k+N)$, and the corresponding received sequence is $Y(k-N:k+N)$. All data in lookup table are set to 0 in the initial state, and the sliding window selects $2N+1$ symbols in the transmission sequence each time, after looking up the index and finding the address of this pattern, the error $E(k)$ is obtained by subtracting the central symbol of the sending sequence and the receiving sequence. As the sliding window moves forward, error values of all transmission modes can be traversed. Suppose the lookup table index is i, the number of data stored in index is $M(i)$, and the lookup table is updated as follows

$$LUT(i) = LUT(i) + E(k) \tag{4}$$

$$M(i) = M(i) + 1 \tag{5}$$

$$LUT(i) = \frac{LUT(i)}{M(i)} \tag{6}$$

3.1.3 Summary

In terms of modulation and impairment suppression technology in transmitter of optical transmission system, researchers have carried out a lot of research, such as peak to average power ratio (PAPR) reduction technology, DAC resolution enhancement [8], joint pre-equalization in electrical and optical domain, etc. In recent years, artificial intelligence techniques have been recently proposed as a promising tool to address various challenges in optical communication, and machine learning technology based on indirect learning [9] and neural network [10] has also been used for impairment compensation of transmitter devices.

3.2 Impairment compensation techniques at receiver side

In the long haul and large capacity optical transmission system, the optical link will introduce chromatic dispersion [11], polarization mode dispersion (PMD) [12], fiber nonlinearity, etc., the laser linewidth and frequency jitter will bring frequency offset and phase noise [13, 14], and the ADC sampling frequency and phase cannot be synchronized with the DAC at the transmitter. All these problems can be solved by using mature DSP technology, so as to avoid the use of a series of complex devices such as phase locked loop. With the development of ASIC chip manufacturing technology, highly integrated and flexible digital signal processing technology can meet the needs of high-speed optical transmission system in the future.

As shown in the **Figure 9**, the basic DSP algorithm flow of a typical coherent optical communication receiver, including IQ imbalance compensation, CD compensation, timing recovery, polarization demultiplexing, frequency offset estimation, and carrier phase recovery. According to the algorithm design, different

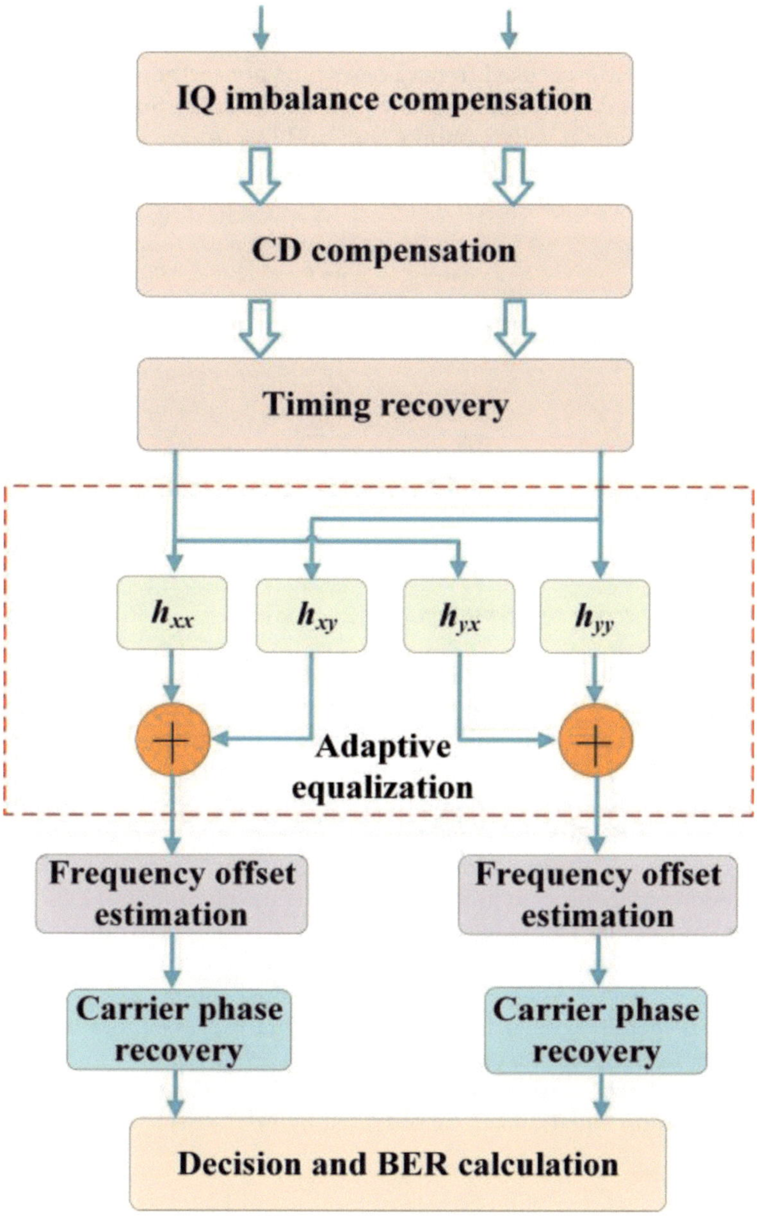

Figure 9.
DSP flow of coherent optical transmission system.

algorithms may have nested or parallel processing, these algorithms are closely linked and indispensable, which is the basis of coherent optical communication with high baud rate and high order modulation format.

3.2.1 I/Q imbalance compensation

Ideally, the I and Q components in the received signal are completely orthogonal, but in the actual system, for the extinction ratio of the two arms of the IQ modulator is not completely consistent, the division ratio of the 3 dB coupler in the receiver is asymmetric, and the response of the balance detector is inconsistent, the amplitude and phase of the two IQ components will be imbalanced. So, it is necessary to implement IQ imbalance compensation and normalization in the first step of digital signal processing. In general, Gram-Schmidt orthogonalization procedure (GSOP) algorithm is used to map one set of non-orthogonal vectors as reference variables and the other set as orthogonal variables. Suppose that $I_{in}(k)$ and $Q_{in}(k)$ are non-orthogonal vectors, and $I_{out}(k)$ and $Q_{out}(k)$ are vectors processed by GSOP, as

$$I_{out}(k) = \frac{I_{in}(k)}{\sqrt{P_I}} \tag{7}$$

$$Q'(k) = Q_{in}(k) - \frac{\rho I_{in}(k)}{P_I} \tag{8}$$

$$Q_{out}(k) = \frac{Q'(k)}{\sqrt{P_Q}} \tag{9}$$

where $\rho = E\{I_{in}(k) \cdot Q_{in}(k)\}, P_I = E\{I_{in}{}^2(k)\}, P_Q = E\left\{Q'^2(k)\right\}$, and $E\{\cdot\}$ represents expectation.

3.2.2 Chromatic dispersion compensation

Chromatic dispersion is a static impairment for optical signal in fiber transmission. The main factor of CD is that the characteristics of optical fiber material lead to different propagation group velocity of different frequency components of optical signal, which is similar to the multipath effect in wireless communication, resulting in time-domain pulse broadening. For the early optical fiber communication system, chromatic dispersion is mainly compensated by negative dispersion coefficient media such as dispersion compensation fiber, fiber Bragg grating and other dispersion compensation modules. With the development of DSP technology, digital signal processing technology can completely replace the function of optical dispersion compensation module, it is easy to realize dispersion compensation based on DSP.

Generally, dispersion coefficient D is used to quantify the pulse broadening caused by fiber dispersion, the unit is $ps/nm/km$. The partial differential equation of the influence of fiber dispersion on signal envelope is derived, and the frequency domain transmission equation can be obtained by Fourier transform

$$G(z, w) = exp\left(-j\frac{D\lambda^2}{4\pi c}\omega^2\right) \tag{10}$$

where λ is the wavelength of light wave, c represents light speed, and ω is arbitrary frequency component. The frequency-domain transfer function of the dispersion compensation filter is obtained by inversing the dispersion coefficient of the transfer function as

$$G(z, w) = exp\left(j\frac{D\lambda^2}{4\pi c}\omega^2\right) \tag{11}$$

In the long-distance optical communication system, the signal sub block must be large enough to compensate for the dispersion effect in the transmission. Therefore, an overlapped frequency domain equalization structure is proposed to improve the transmission and DSP efficiency by forming overlaps between sub blocks as shown in **Figure 10**.

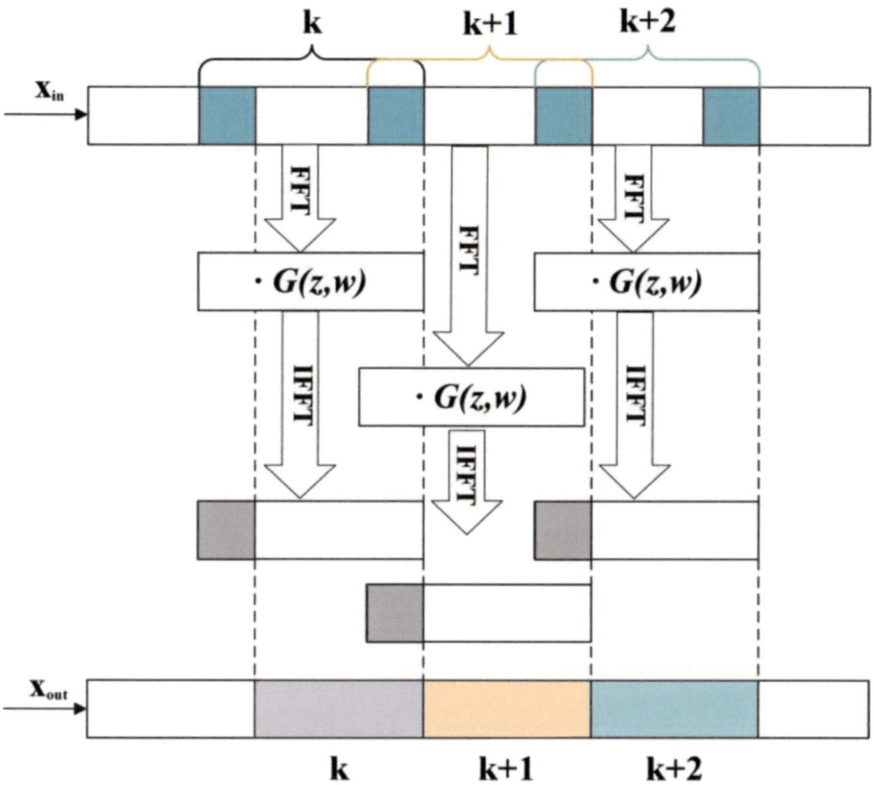

Figure 10.
Frequency domain dispersion equalization with overlap method.

3.2.3 Clock recovery

After photodetection, the electrical signal is sampled and digitized by the ADC. However, in the actual system, because the local sampling clock is not synchronized with the transmitter signal clock, the sampling point of the ADC is not the best sampling point of the signal in most cases. On the other hand, due to the instability of the local clock source itself, it may also cause the sampling error of the system. This sampling error includes both the sampling phase error and the sampling frequency error. The clock error of the sampling signal, on the one hand, is due to the imperfection of the sampling point, causing interference between sampling symbols; on the other hand, the jitter of the sampling clock will also cause fluctuations in signal performance. Therefore, in order to achieve optimal digital signal recovery, a clock recovery module is needed in the actual system to eliminate the impact of clock sampling errors. Considering that the dispersion will cause the disappearance of the clock component, usually, the clock recovery module is placed after the dispersion compensation or works with the dispersion compensation module to form a unified balanced feedback module.

The feedback time-domain clock recovery algorithm is proposed by Gardner [15]. This algorithm uses a feedback clock synchronization structure to estimate the phase of the retiming digital clock source feedback by calculating the timing error. The estimation of the timing error can track the frequency jitter of the signal, and the application of this algorithm can achieve dynamic clock recovery. On the other hand, the Gardner clock recovery algorithm only needs two samples per symbol and the algorithm complexity is low. It is widely used in the digital signal processing module of the coherent optical communication system.

3.2.4 Adaptive equalization and polarization demultiplexing

As CD compensation technology has been well promoted in optical fiber communication systems, PMD becomes the primary impairment which limits the information capacity and transmission distance for over 40Gbit/s systems [16–18]. Besides, PMD is a stochastic impairment that converts with time, temperature, wavelength and fiber conditions which makes PMD hard to estimate and compensate. In recent years, the research on how to overcome the performance deterioration of optical communication system caused by PMD effect has been a hot topic, most of which focuses on PMD compensation [19, 20].

The adaptive equalizer is generally used to polarization demultiplex and PMD compensation of channels. A two-by-two multiple-input multiple-output (MIMO) structured finite impulse response (FIR) filters as shown in **Figure 11** is used to estimate the inverse Jones matrix of the dynamic channel [17].

The input sequence of the filter is symbol-spaced with index n, while the N tap FIR filters, h_{xx}, h_{xy}, h_{yx} and h_{yy} are the column vector of length N. Tap weights are updated every two samples as the input sequence is two-fold sampled. Therefore, x_i and y_i represent a sliding block of N samples such that

$$x_i(n) = [x_i(n), x_i(n-1) \dots x_i(n-N)] \tag{12}$$

$$y_i(n) = [y_i(n), y_i(n-1) \dots y_i(n-N)] \tag{13}$$

We consider that $u_i(\mathrm{n}) = [x_i(n), y_i(n)]$, $h_x(\mathrm{n}) = [h_{xx}(n), h_{xy}(n)]$, $h_y(\mathrm{n}) = [h_{yx}(n), h_{yy}(n)]$. And the filters outputs form as

$$x_o(n) = h_x^H(n) u_i(n) \tag{14}$$

$$y_o(n) = h_y^H(n) u_i(n) \tag{15}$$

where superscript$(.)^H$ means the conjugate transpose.

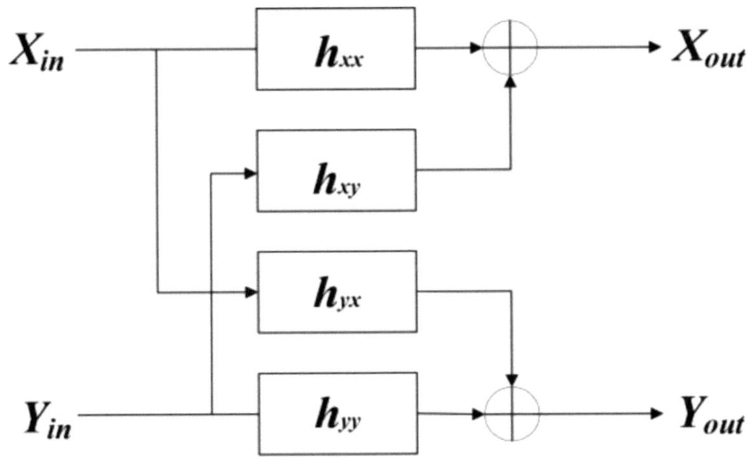

Figure 11.
Framework of 2 × 2 MIMO structured FIR filters.

For fast adaptive equalizer, we generally use stochastic gradient descent (SGD) optimizer to update the tap weights. Meanwhile, we need to choose a cost function to describe the error degree of the output samples so that the equalizer can get the response and update tap weights. For constant modulus algorithm (CMA), its cost functions are given as

$$\varepsilon_x(n) = R_2 - |x_o(n)|^2 \tag{16}$$

$$\varepsilon_y(n) = R_2 - |y_o(n)|^2 \tag{17}$$

where R_2 is the real-valued constant and given by $R_2 = E|X_{sym.}|^4/E|X_{sym.}|^2$. The update equations of FIR filters are given as

$$h_x(n+1) = h_x(n) + \mu\varepsilon_x(n)x_o^*(n)u_i(n) \tag{18}$$

$$h_y(n+1) = h_y(n) + \mu\varepsilon_y(n)y_o^*(n)u_i(n) \tag{19}$$

where μ is the step size parameter and the superscript$(.)^*$ means the complex conjugate operation.

The cost functions of CMA describe error degree between the amplitude of output symbols and the proposed convergence radius [17, 18]. By updating the tap weights of FIR filters with cost functions, the error degree can be minimized in a certain extent and the output symbols can converge on a circle with the proposed radius that equals to $\sqrt{R_2}$. The value of the constant R_2 and step size can significantly affect the convergence degree of CMA. Appropriate R_2 helps CMA to converge in less steps. Shorter step size can help CMA to get better convergence, but the computation of the algorithm is also increased. CMA needs enough steps to update and optimize its FIR filters. The output performance of CMA will be insufficient if the proposed convergence length is not long enough.

All tap weights are initialized to zero except the central tap of h_{xx} and h_{yy}, which are initialized to unity. The filter taps of h_{xx}, h_{xy}, h_{yx} and h_{yy} estimate the components for data sequences. At the transmitter, data sequences for X and Y polarization are independent and only contain their own information. Thus, the central tap of h_{xx} and h_{yy} are set to unity. After fiber transmission, the received signal is contaminated by channel impairments and noise. With the contamination of noise and the channel impairments caused by linear effects such as CD, PMD etc. or fiber nonlinear effects, the received data sequences for X and Y polarization are no longer independent. One or more symbols in the received sequences can interfere other symbols, while the symbols for one polarization can interfere other symbols for another polarization. The interference between polarizations can be estimated by filter taps of h_{xy} and h_{yx}. Though updating filter tap weights, the 2×2 MIMO structured FIR filters can gradually identify the components of each data sequence, and the dynamic impairments of channel can be compensated. This process can also achieve polarization demultiplexing.

It is worth noting that the CMA can be implemented in a full-blind mode, but it set no constrain with its outputs. Therefore, it is possible for the equalizer to converge on the same output, corresponding to the Jones matrix becoming singular. In practice, we need to check if h_x and h_y become singular after the algorithm running for certain steps, and if so, a mathematic process is necessary to make h_x and h_y nonsingular and the whole algorithm should be restarted.

CMA is especially suited to the modulation format with constant amplitude such as quadrature phase-shift keying (QPSK) and M-ary phase-shift keying (MPSK). For the formats with inconstant amplitudes such as quadrature amplitude modulation, the CMA error cannot converge to zero as extra noise is introduced during equalizing. To improve the SNR performance for the formats with inconstant amplitudes, several multi-modulus algorithms such as radius-directed algorithm (RDA) and cascaded multi-modulus algorithm (CMMA) are established [21, 22].

As CMA set at real-valued constant R_2, CMMA set several constants which depends on the ideal constellations. According to the distance between the input symbol and the constellation origin, the algorithm estimates which circle the symbol belongs to. Then CMMA calculates the error using the estimated radius. The rest of CMMA algorithm steps are same as CMA.

Compared with CMA, the CMMA significantly improves the SNR performance for high order QAM. But it reduces the robustness of the filter converging process. This is because multi-modulus algorithm depends on correct decision on symbol radius. For high order QAM format, the distance between different circles is less than the minimum symbol interval. Therefore, the algorithm may make massive mistake over the circle decision if the signal is severely contaminated by channel impairments and noise.

3.2.5 Frequency offset estimation

In the coherent optical communication system, the transmitting laser and the local oscillator laser work independently, so the central wavelength cannot be exactly the same, so there is a certain frequency deviation Δf. It will introduce a continuous phase variation along with time to the received signal, resulting in constellation rotation, so it is necessary to estimate and compensate the frequency offset by DSP. Through FFT operation, the spectrum of received symbol's fourth power value is obtained and analyzed, it can be found that there is a peak component at the frequency of $4\Delta f$, therefore, Δf can be obtained by searching for the maximum spectral component of the fourth-power value of received symbol. Usually, due to the lack of spectral resolution and other reasons, there will be residual frequency offset after estimation, which can be looked at as additional phase noise and recovered by carrier phase recovery.

3.2.6 Carrier phase recovery

Carrier phase recovery (CPR) is an essential DSP unit in coherent systems and has been extensively investigated for QPSK and QAM signals. Like frequency offset estimation algorithms, carrier recovery algorithms can be classified as either blind or data-aided estimation techniques. And the algorithms can be implemented in feedforward manner or in feedback structure [23, 24]. Compared to the constellation for QPSK, the constellation points for QAM vary in both phase and amplitude. Moreover, the modulated signal phase is with multiple values. Viterbi-Viterbi phase estimation (VVPE) algorithm using a fourth-power operation which is suitable for QPSK is hard to completely remove the signal phase and estimate the carrier phase. At present, the algorithms for QAM CPR mainly include Blind phase searching (BPS), improved BPS (BPS/maximum likelihood), decision aided maximum likelihood (DA-ML), etc. [25–27]

BPS is recognized as a favorable solution due to its high performance and suitability for parallel processing and can be used in feedforward manner.

Figure 12 shows the block diagram of the BPS algorithm in pure feedforward manner. The input signal x_i is sampled at the symbol rate. The received signal x_i is rotated by B test carrier angles φ_b with

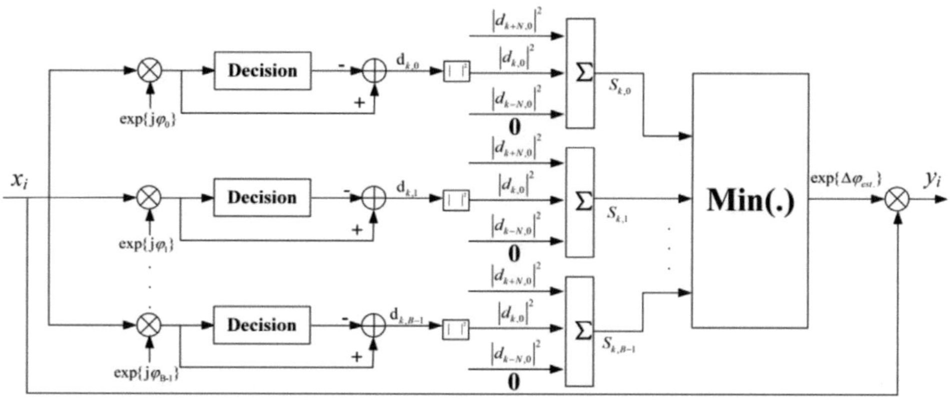

Figure 12.
BPS diagram in feedforward manner.

$$\varphi_b = \frac{b}{B} \bullet \frac{\pi}{2}, b \in \{0, 1, \ldots, B-1\} \tag{20}$$

then all rotated symbols are fed into a decision circuit, which output the ideal constellation points with the minimum Euclidean distance to the input symbols. The squared distance $|d_{k,b}|^2$ to the closest constellation point is calculated in the complex plane

$$|d_{k,b}|^2 = \left|x_k \exp\{j\varphi_b\} - \lfloor x_k \exp\{j\varphi_b\}_D \rfloor\right|^2 \tag{21}$$

in order to remove the other initial noise distortions from the receiver, the distance of 2N + 1 consecutive test symbols rotated by the same carrier phase angle φ_b are summed up

$$s_{k,b} = \sum_{n=-N}^{N} |d_{k-n,b}|^2 \tag{22}$$

the optimum value of the filter half width N depends on the laser linewidth times symbol rate product. N = 6, …,10 is generally a good choice.

After filtering the optimum phase angle $\Delta\varphi_{est.}$ by searching the minimum sum of distance values, then the output symbols y_i, which is the input symbols x_i rotated by $\Delta\varphi_{est.}$, is outputted for the following DSP in coherent systems.

Due to the 4-fold ambiguity of the recovered phase in the square M-QAM, the blind algorithms may cause incorrect phase estimation by a multiple of $\pi/2$ causing cycle slip. This problem can be solved by using framing information or by applying differential coding [23–25].

Though BPS shows a good tolerance to laser phase noise and can be flexibly applied to higher order QAM, with an increasing modulation order a larger number of test phases are required and the computation complexity increases. Therefore, an improved BPS algorithm with a two-stage diagram has been established [24]. The first stage of improved BPS just requires rough estimation and the required number of test phase φ_b can be reduced. A maximum likelihood phase estimator is introduced in the second phase to improve the accuracy. This two-stage BPS/ML algorithm effectively improves the performance with the computation complexity and availability of BPS algorithm remained.

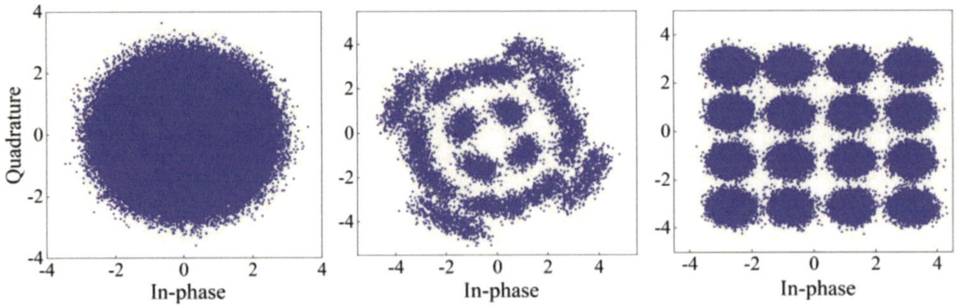

Figure 13.
Constellations after CD compensation, polarization demultiplexing, and carrier phase recovery.

3.2.7 Summary

This section introduces a series of impairment compensation algorithms for coherent optical communication system. Here, the constellation diagrams of 60GBaud 16QAM signal in DSP process is given in the **Figure 13**. Through the algorithms in this section, we can realize the signal orthogonalization and normalization, compensate the fiber dispersion, eliminate the clock sampling error, depolarize the multiplexing and equalize channel response at the same time, and finally recover the carrier phase to get the constellation of the original transmission signal. A series of impairment compensation algorithms based on DSP at the receiver end lay the foundation for high-speed, large-capacity and long-haul optical transmission.

4. Conclusion

In this chapter, first we propose a probabilistic shaping 16QAM modulation scheme based on trellis coded modulation. Through non-uniform probability mapping of TCM-16QAM subset, an effective and good overall probability distribution of 16QAM constellation is obtained. The scheme is successfully demonstrated in a 25 km single-mode fiber transmission system, and better OSNR gain and BER performance are obtained. Then impairment compensation techniques in coherent optical communication are introduced from two aspects, transmitter side and receiver side, pre-emphasis and look-up table technology have been widely used in the impairment compensation of the transmitting devices, while the DSP process at the receiver side is more complicated, including GSOP, clock recovery, dispersion compensation, dynamic equalization, and carrier phase recovery. Using DSP technology can effectively mitigate the impairment of optoelectronic devices and optical fiber links, and it is an effective method to realize ultra-high speed, large-capacity and long-haul transmission.

Acknowledgements

This work is supported by National Natural Science Foundation of China No.62022016, No.61835002, No.61727817, National Key R&D Program of China from Ministry of Science and Technology, China No.2019YFA0706300, Open Fund of IPOC (BUPT) No.IPOC2020A007, No.IPOC2020A006, and China Postdoctoral Science Foundation No.2020 M680384, No.2020 M680385.

Author details

Zhipei Li, Dong Guo and Ran Gao*
School of Information and Electronics, Beijing Institute of Technology, Beijing, China

*Address all correspondence to: 6120190142@bit.edu.cn

References

[1] Yankov M P, Zibar D, Larsen K J, Christensen L, Forchhammer S. Constellation Shaping for Fiber-Optic Channels with QAM and High Spectral Efficiency. IEEE Photonics Technology Letters, 2014; 26(23):2407-2410. DOI: 10.1109/LPT.2014.2358274

[2] Gui T, Wang X, Tang M, Yu Y, Li L. Real-Time Demonstration of 600-Gb/s DP-64QAM Self-Homodyne Coherent Bi-Direction Transmission with Un-Cooled DFB Laser. Optical Fiber Communication Conference. 2020

[3] Napoli A, Sun H, Maher R, Torbatian M, & Zhang J, 800G DSP ASIC Design Using Probabilistic Shaping and Digital Sub-Carrier Multiplexing. Journal of Lightwave Technology, 2020; 38(17): 4744-4756. DOI: 10.1109/JLT.2020.2996188

[4] Zhou X, Urata R, Liu H. Beyond 1Tb/s Datacenter Interconnect Technology: Challenges and Solutions. Optical Fiber Communication Conference. 2019.

[5] Raphaeli D, Gurevitz A, Constellation shaping for pragmatic turbo-coded modulation with high spectral efficiency. IEEE Transactions on Communications, 2004; 52, 341–345. DOI: 10.1109/tcomm.2004.823564

[6] Zhang Z, Li C, Chen J, Ding T, Wang Y, Xiang H, Xiao Z, Li L, Si M, Cui X. Coherent transceiver operating at 61-Gbaud/s. Optics Express, 2015; 23 (15):18988-95. DOI: 10.1364/OE.23.018988

[7] Napoli A, Berenguer P W, Rahman T, Khanna G, Mezghanni M M, Gardian L, Riccardi E, Piat A C, Calabrò S, Dris S. Digital pre-compensation techniques enabling high-capacity bandwidth variable transponders. Optics Communications, 2017; S0030401817308465. DOI: 10.1016/j.optcom.2017.09.062

[8] Sadot D, Yoffe Y, H Faig, Wohlgemuth E. Digital Pre-Compensation Techniques Enabling Cost-Effective High-Order Modulation Formats Transmission. Journal of Lightwave Technology, 2018; 37(2): 441-450. DOI: 10.1109/JLT.2018.2888941

[9] Paryanti G, Faig H, Rokach S L, Dan S. A Direct Learning Approach for Neural Network Based Pre-Distortion for Coherent Nonlinear Optical Transmitter. Journal of Lightwave Technology, 2020; 38(15): 3883-3896. DOI: 10.1109/JLT.2020.2983229

[10] Freire P J, Neskorniuk V, Napoli A, Spinnler B, Turitsyn S. Complex-Valued Neural Network Design for Mitigation of Signal Distortions in Optical Links. Journal of Lightwave Technology, 2021; 39(6):1696-1705. DOI: 10.1109/JLT.2020.3042414

[11] Prince M. Performance Limitations of an Optical TDM Transmission Link Caused by Fiber Chromatic Dispersion. 2020 FORTEI-International Conference on Electrical Engineering (FORTEI-ICEE); 2020. DOI:10.1109/FORTEI-ICEE50915.2020.9249807

[12] Ip E, Kahn J M. Digital Equalization of Chromatic Dispersion and Polarization Mode Dispersion. Journal of Lightwave Technology, 2007; 25(8): 2033-2043. DOI: 10.1109/JLT.2007.900889

[13] Huerta-Cuellar G, Pisarchik A N, Barmenkov Y O. Experimental characterization of hopping dynamics in a multistable fiber laser. Physical Review E, 2008; 78(3):035202. DOI: 10.1103/PhysRevE.78.035202

[14] Fatemi F K, Lou J W, Carruthers T F. Frequency comb linewidth of an actively mode-locked fiber laser. Optics Letters, 2004; 29(9):944-946. DOI: 10.1364/OL.29.000944

[15] Gardner F M. A BPSK/QPSK timing-error detector for sampled receivers. IEEE Transactions on Communications, 1986; 34(5):423-429. DOI: 10.1109/TCOM.1986.1096561

[16] Kikuchi K, Fundamentals of Coherent Optical Fiber Communications. Journal of Lightwave Technology, 2016; 34(1):157-179. DOI: 10.1109/JLT.2015.2463719.

[17] Faruk M S, Savory S J. Digital Signal Processing for Coherent Transceivers Employing Multilevel Formats. Journal of Lightwave Technology, 2017, 35(5): 1125-1141. DOI:10.1109/JLT.2017.2662319

[18] Lagha M K, Gerzaguet R, Bramerie L, Gay M, Scalart P, Blind Joint Polarization Demultiplexing and IQ Imbalance Compensation for *M* -QAM Coherent Optical Communications. Journal of Lightwave Technology. 2020, 38, 4213-4220. DOI: 10.1109/JLT.2020.2986601

[19] El-Fiky E, Chagnon M, Sowailem M, Samani A, Morsy-Osman M, Plant D V. 168-Gb/s Single Carrier PAM4 Transmission for Intra-Data Center Optical Interconnects. IEEE Photonics Technology Letters, 2017; 29(3):314-317 DOI: 10.1109/LPT.2016.2647232.

[20] Yu Z, Zhao Y, Hu S, Wan Z, Xu K. PDL and CD insensitive low complexity equalizer for short reach coherent systems. Optics Express, 2021; 29(5), 6657-6667. DOI: 10.1364/OE.418456

[21] Zhang T, Xiang Q, Zhang S, Liu L, & Zuo T. Cost-effective digital coherent short-reach transmission system with D8QAM and low-complexity DSP. Optics Express, 2021; 29, 11892-11902. DOI: 10.1364/OE.422456

[22] Zhang Q, Li X, Zhang N, Hu L, Xi L, A Low Complexity Frequency Domain Adaptive Equalizer for Coherent Optical Receivers. Asia Communications and Photonics Conference/International Conference on Information Photonics and Optical Communications 2020 (ACP/IPOC), 24-27 October 2020; Beijing, China. M4A.287.

[23] Pfau T, Hoffmann S, R. Noe. Hardware-Efficient Coherent Digital Receiver Concept with Feedforward Carrier Recovery for MM -QAM Constellations. Journal of Lightwave Technology, 2009; 27(8): 989-999, DOI: 10.1109/JLT.2008.2010511

[24] X. Zhou. An Improved Feed-Forward Carrier Recovery Algorithm for Coherent Receivers With M-QAM Modulation Format. IEEE Photonics Technology Letters, 2010; 22(14): 1051-1053. DOI: 10.1109/LPT.2010.2049644.

[25] Zhao J, Chen L K. Carrier Phase Recovery Based on KL Divergence in Probabilistically Shaped Coherent Systems. Journal of Lightwave Technology. 2021; 39, 2684-2695. DOI: 10.1109/JLT.2021.3054949

[26] Borjeson E, Fougstedt C, Larsson-Edefors P. VLSI Implementations of Carrier Phase Recovery Algorithms for M-QAM Fiber-Optic Systems. Journal of Lightwave Technology. 2020: 38, 3616-3623. DOI: 10.1109/JLT.2020.2976166

[27] Lu J, Fu S, Hu Z, D Lei, Tang M, D Liu. Carrier Phase Recovery for Set-Partitioning QAM Formats. Journal of Lightwave Technology, 2018; 36(18): 4129-4137. DOI: 10.1109/JLT.2018.2859238

Chapter 5

OAM Modes in Optical Fibers for Next Generation Space Division Multiplexing (SDM) Systems

Alaaeddine Rjeb, Habib Fathallah and Mohsen Machhout

Abstract

Due to the renewed demand on data bandwidth imposed by the upcoming capacity crunch, optical communication (research and industry) community has oriented their effort to space division multiplexing (SDM) and particularly to mode division multiplexing (MDM). This is based on separate/independent and orthogonal spatial modes of optical fiber as data carriers along optical fiber. Orbital Angular Momentum (OAM) is one of the variants of MDM that showed promising features including the efficient enhancement of capacity transmission from Tbit to Pbit and substantial improvement of spectral efficiency up to hundreds (bs^{-1} Hz^{-1}). In this chapter, we review the potentials of harnessing SDM as a promising solution for next generation global communications systems. We focus on different SDM approaches and we address specifically the MDM (different modes in optical fiber). Finally, we highlight the recent main works and achievements that have been conducted (in last decade) in OAM-MDM over optical fibers. We focus on main R&D activities incorporating specialty fibers that have been proposed, designed and demonstrating in order to handle appropriates OAM modes.

Keywords: Space Division Multiplexing (SDM), Mode Division Multiplexing (MDM), Orbital Angular Momentum (OAM), Specialty optical fibers

1. Introduction

Bandwidth-hungry applications and services, such as HDTV, big data, quantum computing, 5G/6G communication, industry 4.0 and game streaming, in addition to the exponential increase of users and connected devices (Internet of Things: IOT), may cause a capacity crunch in near future [1–3]. While other physical limitations behind the capacity crunch are based on the nonlinear Shannon limit and the scalability of actually deployed devices. The cited emerging applications (i.e. paradigms) has pushed telecommunications community (researchers & industries) to grow through multiple stages by developing higher capacity optical networks in optical fiber based links targeting to deal with the evolution of the market need for telecoms and Internet data services and paving the road to surpass the upcoming capacity limit challenges [4].

Recently, the capacity and the spectral efficiency of optical fibers have been substantially improved (i.e. scaling by several orders of magnitude) by using different multiplexing techniques and advanced optical modulation formats.

These multiplexing techniques are based on the exploitation of degrees of freedom of the optical signal to encode data information. The time, as time division multiplexing (TDM: interleaving channels temporally), the polarization, Polarization division multiplexing (PDM), the wavelength, as wavelength division multiplexing (WDM: using multiple wavelength channels) and the phase (quadrature) are examples of such techniques [5].

Research and industrial community had recently oriented their effort towards Space Division Multiplexing (SDM) techniques that is based on the exploitation of the spatial structure of the light or the physical transmission medium to encode information. Simply, SDM consists of increasing the number of data channels available inside an optical fiber. Two attractive embodiments of SDM are core division multiplexing (CDM) and mode division multiplexing (MDM) [6]. CDM is simply considered as the increasing of parallel single mode cores, carrying information, embedded in the same cladding of optical fiber (known as multicore fiber MCF) or single core fibers bundles [7]. Mode division multiplexing (MDM) is based on excitation and propagation of several spatial optical modes as individual/separate/independent data channels within common physical transmission medium targeting to boost the capacity transmission [8]. MDM is realized by multimode fibers generally over short haul interconnect transmission or few mode fibers as transmission medium for long haul transmission link. Numerous mode basis have been used for mode division multiplexing showing its effectiveness to scale up from Terabit to Petabit the capacity transmission and unleash from dozen to hundred (bit/s/Hz) the spectral efficiency over optical fiber.

It is well known that light can carry Angular Momentum (AM) that expresses the amount of dynamical rotation present in the electromagnetic field representing the light. The AM of light beam is divided into two distinct forms of rotation: Spin Angular Momentum (SAM) and Orbital Angular Momentum (OAM) [9]. The SAM is related to the polarization of light (e.g. right or left in circular polarization) while the OAM is related to the spiral phase front of exp. (jlφ) where l is a topological charge number (arbitrary unlimited integer), and φ is the azimuthal angle. Orbital angular momentum (OAM) of light, (known as twisted light), an additional degree of freedom, is arguably one of the most promising approaches that has recently deserved a special attention in optical fiber networks. Benefiting from two inherent features, which are:

(1) The orthogonality: where as a definition two signals are orthogonal, if data sent in these two dimensions can be uniquely separated from one another at the receiver without affecting each other's detection performance. Two OAM modes with different charge number l do not interfere.

(2) The unlimitleness: the charge number l is theoretically infinite. Hence, Each OAM mode (each specifically l) is an independent data channels. OAM modes has been harnessed in multiplexing/de-multiplexing (OAM-MDM) or in increasing the overall optical channel capacity [10, 11].

As any promising technology, OAM-MDM through optical fibers is facing several key challenges, and lots crucial issues that it is of great importance to handle with it in order to truly realize the full potential of this technique and to paving the road to a robust transmission operation with raised performances in future communication systems.

In strict sense 'Mode division multiplexing', means that the modes (channels) are separate and should remain uncoupled and not interfere with each other (i.e. orthogonal). Hence, mode coupling (e.g. channels crosstalk) is the major obstacle for OAM-MDM. Channels crosstalk is obviated by either fiber design or multiple input multiple output digital signal processing (MIMO DSP) [9–11].

By carefully manipulating the fiber design parameters, it is possible to supervise the interactions between propagated modes and even control which modal basis is incorporated: LP-fibers where the separation between vector modes are inferior to $1 \times 10\text{–}4$, or OAM-fibers where the intermodal separation exceeds $1 \times 10\text{–}4$, since either LP or OAM modes are constructed from fiber eigenmodes themselves [9]. This better facilitates understanding each fiber parameter impact and smooth the way of transition from design stage to fabrication process. Adding to that, exploit MIMO DSP is considered as the extreme choice to decipher channels at the receiving stage since it is heavy and complex. Its complexity is came from its direct proportionality to the transmission distance and to the number of modes. This allow it to become impractical in real time and threats the scalability of MDM in next generation optical communication system. For OAM-MDM systems using optical fibers, the fiber design stage is considered as the most crucial part and there is still a lot of opportunities for improved designs. New fiber designs for OAM mode transmission over short/medium and longer distances or among higher number of modes or possess a high performance metrics have been proposed and examined.

With the different related key challenges, this chapter offers a review of the state-of-the-art of SDM advances especially on OAM-MDM over optical fibers. In the first section, we discuss the SDM approaches as a solution to the expected capacity limit. The different mode basis supported in optical fibers are presented and discussed as either cylindrical vector modes, LP modes or OAM modes. The second section acts as a survey on recent advances (over last ten years) in OAM-MDM over optical fibers. We review the research effort invested in harnessing OAM as a degree of freedom to carry data in optical fiber networks. We summarized the key obtained results in the main family of optical fibers (i.e. conventional fibers and OAM specialty fibers) using OAM modes.

2. SDM over optical fibers

Space division multiplexing (SDM) has attracted high interest. It has revealed multiple directions of exploration and development. SDM consists of exploiting space-independent communication channels in both guided waves (e.g. optical fibers) or free space optical link (FSO). The channels' type vary depending in which factor of SDM we are exploiting; diversified cores, multiplexed LP modes or modes carrying OAM, multiple cores each supporting few multiplexed LP modes and so on.

Two main subset in SDM could be explored: core division multiplexing (CDM) where information is transmitted through cores (or fibers) of multicore fibers or mode division multiplexing (MDM), where information is transmitted through propagating modes of few or multimode fibers.

2.1 Core division multiplexing (CDM)

In principle, two main schemes are used. The first is based on the use of Single-core Fiber bundle (i.e. fiber ribbon) where parallels single mode fibers are packed together creating a fiber bundle or ribbon cable. The overall diameter of these bundles varies from around 10 mm to 27 mm. Fiber bundles deliver up to hundreds of parallel links. Fiber bundles have been commercially available [12, 13] and deployed in current optical infrastructure for several years already. Fiber bundles are also commercially used in conjunction with several SDM transceiver technologies [14].

The second scheme is based on carrying data on single cores (each core supports single mode) embedded in the same fiber known as Multicore Fibers (MCFs). Hence, each core is considered as an independent single channel (**Figure 1**).

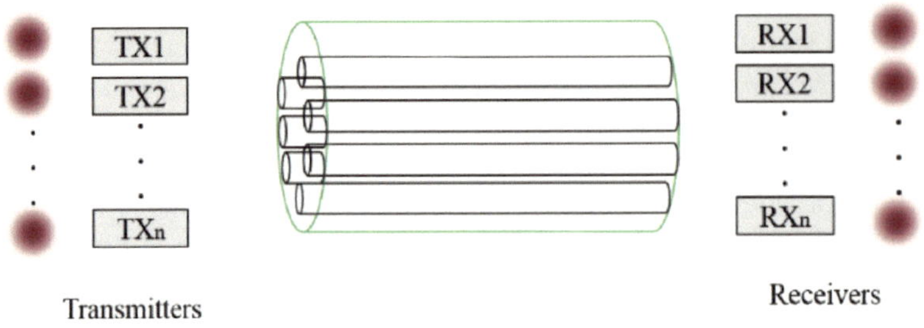

Figure 1.
SDM through MCF.

The most important constraint in MCFs is the inter-core crosstalk (XT) caused by signal power leakage from core to its adjacent cores that is controlled by core pitch (distance between adjacent cores denoted usually as Λ) [15]. There are in Principle, two main categories of MCF: weakly coupled MCFs (=uncoupled MCF) and strongly coupled MCFs (=coupled MCF) depending on the value of a coupling coefficient 'K' (used to characterize the intercore crosstalk) [16–18]. Using the so-called supermodes to carry data, the crosstalk in coupled MCF must be mitigated by complex digital signal processing algorithms, such as multiple-input multiple-output digital signal processing (MIMO-DSP) techniques [19]. On the contrary, due to low XT in uncoupled MCF, it is not necessary to mitigate the XT impacts via complex MIMO. In principle, three crosstalk suppression schemes in uncoupled MCF could be incorporated, which are trench-assisted structure, heterogeneous core arrangement, and propagation-direction interleaving (PDI) technique [7].

The first paper on communication using MCF demonstrate a transmission of 112-Tb/s over 76.8 km in a 7-cores fiber using SDM and dense WDM in the C + L ITU-T bands. The spectral efficiency was of 14 b/s/Hz [20]. The second paper [21] shows an ultra-low crosstalk level (≤ -55 dB over 17.6 km), which presents the lowest crosstalk between neighboring cores value to date. Other reported works, show high capacity (1.01Pb/s) [22] over 52 km single span of 12- core MCF. In [23], over 7326 km, a record of 140.7 Tb/s capacity are reached.

2.2 Mode division multiplexing (MDM)

Carrying data on optical fiber modes known as mode division multiplexing. In that scenario, each propagating mode is considered as independent channel [5, 24]. Two types of fiber are dedicated to support that strategy. One is based on the use of multimode fibers (MMF) while the second exploits the known few-mode fibers (FMF). The main difference between both is the number of modes (available channels). Since MMF can support large number of modes (tens), the intermodal crosstalk becomes large as well as the differential mode group delay (DMGD), where each mode has its own velocity, reducing the number of propagating modes along the fiber becomes viable solution. This supports FMF as a viable candidate for realizing SDM [5]. The concept of mode division multiplexing over a few/multi-mode fiber is illustrated in **Figure 2**.

Other kinds of optical fiber that can be used in MDM such as photonic crystal fibers (PCFs). Based on the properties of photonic crystals, PCF confines light by band gap effects, using air holes in their cross-sections, or by a conventional higher-index core modified by the presence of air holes. The PCF is built of one material (SiO_2, As_2S_3, Polymers, etc), and air holes are introduced in the area surrounding the core providing the change of the refractive index contrast between the core and

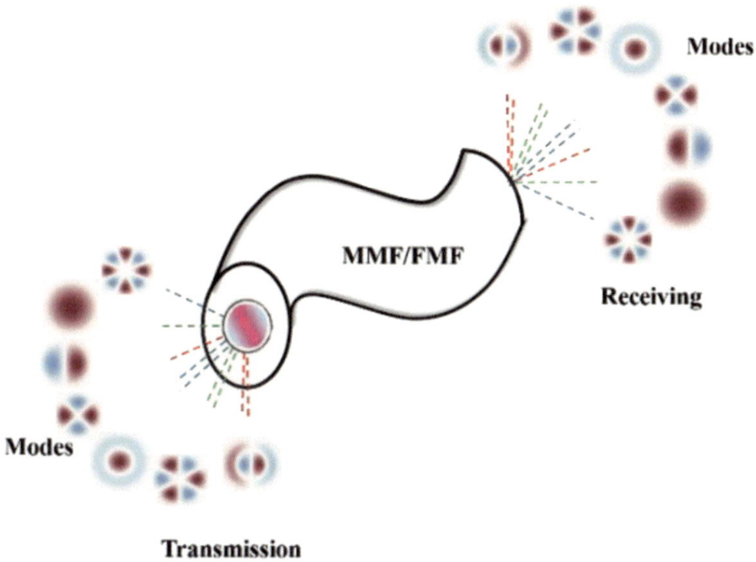

Figure 2.
The concept of mode division multiplexing over a FMF/MMF.

the cladding. The transposition of air holes laid to form a hexagonal or circular lattice. **Figure 3** recapitalizes the principle SDM approaches over optical fibers [25].

2.3 Guided modes of optical fibers

We look into the different modal basis that can be supported by optical fibers. Like all electromagnetic phenomena, the propagation of optical fields along optical fiber is governed by Maxwell's equations. Several modal basis can describe the

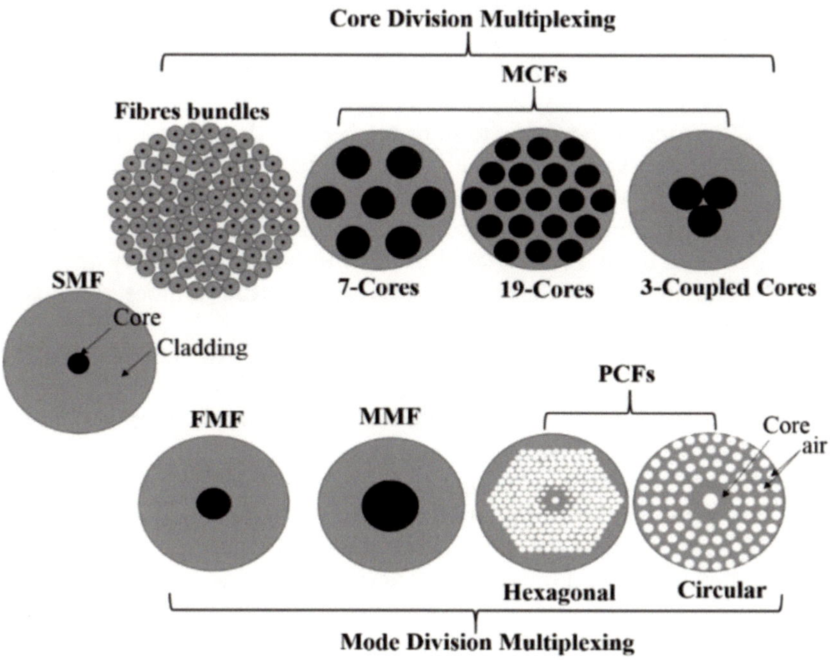

Figure 3.
Different approaches for SDM over optical fibers.

propagation in optical fibers. In this chapter, fiber guided modes that we will meet are vector modes (i.e. fiber eigenmodes), linear polarized modes (i.e. LP modes) and orbital angular momentum modes (i.e. OAM modes). In the following, we provide general notions including mathematical expressions of modes of each mode basis.

2.3.1 Cylindrical vector modes

In the absence of the current in the medium, Maxwell equations are reduced to two homogeneous vector wave equations given by the following expressions [26]:

$$\left(\overrightarrow{\nabla^2} + k^2n^2\right)\vec{E} = -\vec{\nabla}\left(\vec{E}.\vec{\nabla}\ln n^2\right) \tag{1}$$

$$\left(\overrightarrow{\nabla^2} + k^2n^2\right)\vec{H} = \left(\vec{\nabla} \times \vec{H}\right) \times \vec{\nabla}\ln n^2 \tag{2}$$

Where $\vec{E}$ and $\vec{H}$ are the electric and magnetic field respectively and n is the refractive index profile function. If we apply the boundary conditions according to the geometry and fiber refractive index, we get eigenvalues equation. Each solution of that equation is guided mode known by effective index neff. In cylindrical coordinates, for example, the electrical and magnetic fields are expressed as:

$$\begin{cases} \vec{E} = \left[\vec{r}E_r + \vec{\phi}E_\phi + \vec{z}E_z\right]\exp\left(j\beta z - j\omega t\right) \\ \vec{H} = \left[\vec{r}H_r + \vec{\phi}H_\phi + \vec{r}H_z\right]\exp\left(j\beta z - j\omega t\right) \end{cases} \tag{3}$$

Where Er, Hr are radial components, Eϕ and Hϕ are azimuthal components. $\vec{r}$, $\vec{\phi}$ and $\vec{z}$ are unitary vectors. $\beta = 2\pi n_{eff}/\lambda$ is the propagation constant of guided mode, $\omega = 2\pi c/\lambda = kc$ is the pulsation; λ and c are the wavelength and light velocity both in vacuum, respectively. Guided modes in circularly symmetrical optical fiber are denoted as transverse electric ($TE_{0,m}$) or transverse magnetic modes ($TM_{0,m}$), if $E_z = 0$ or $H_z = 0$ respectively. Other kind of modes are $HE_{\nu,m}$ and $EH_{\nu,m}$ those where $E_z \neq 0$ or $H_z \neq 0$ (transverse components) are noted as hybrid modes. The designation $HE_{\nu,m}$ stands for a hybrid mode for which Hz is dominant compared to E_z, while for $EH_{\nu,m}$, E_z is dominant compared to H_z. The indexes ν and m are the azimuthal and radial indices. ν is related to the number of symmetry axes in the azimuthal dependency of the fields, and m is related to the number of zeros in the radial dependency of the fields.

Because of the circular symmetry, the field must keep the same value after a full 2π azimuthal rotation, thus, the components E_z and H_z have a dependency according to $cos(\nu\phi)$ or $sin(\nu\phi)$. hence, in circularly symmetrical optical fiber, hybrid modes are composed by two modes: one *even* while the other is *odd*. In the even mode, the radials components (E_r) and azimuthal component (H_ϕ) are with $cos(\nu\phi)$ (i.e. Ox symmetry). The components E_ϕ and H_r have dependency according to $sin(\nu\phi)$ (i.e. Oy symmetry). The radial, azimuthal and longitudinal electrical field components of even and odd modes are given by the following expressions:

$$\begin{cases} E_r^{even} = e_r(r)\cos\left(\nu\phi\right) \\ E_\phi^{even} = -e_\phi(r)\sin\left(\nu\phi\right) \\ E_z^{even} = e_z(r)\cos\left(\nu\phi\right) \end{cases} \tag{4}$$

$$\begin{cases} E_r^{odd} = e_r(r)\sin(\nu\phi) \\ E_\phi^{odd} = e_\phi(r)\cos(\nu\phi) \\ E_z^{odd} = e_z(r)\sin(\nu\phi) \end{cases} \tag{5}$$

The modes $HE_{\nu,m}^{even/odd}, HE_{\nu,m}^{even/odd}, TE_{0,m}$ *and* $TM_{0,m}$ are usually denoted as vector modes, cylindrical vector modes or fiber eigenmodes.

2.3.2 Scalar modes: LP modes

Frequently, the refractive index difference between core and cladding in optical fiber is very small ($n_{core} \approx n_{cladding}$). We are then under the weakly guiding condition, and some approximations can be applied. The term "$\nabla \ln n^2$" is neglected in expression 1. The wave equation becomes scalar. The resulted modes are linearly polarized designated usually as LPlm modes. LP modes are quasi-TEM guided modes, and have negligible E_z and H_z components. Therefore, they only have one component in the E field and one component in the H field (by convention, either E_x and H_y, or E_y and H_x in cartesien coordinates). This is why we call them scalar modes. The even modes are with $\cos(l\phi)$, while odd modes are varies with $\sin(l\phi)$. l is the azimuthal number while m has the same definition as in vector modes [26]. The electric field components of even and odd modes (after variable separation: radial and azimuthal) are given by the next expressions:

$$E_x^{even} = e_x(r)\cos(l\phi) \tag{6}$$

$$E_y^{even} = e_y(r)\cos(l\phi) \tag{7}$$

$$E_x^{odd} = e_x(r)\sin(l\phi) \tag{8}$$

$$E_y^{odd} = e_y(r)\sin(l\phi) \tag{9}$$

Practically, the LP modes come from linear combination between cylindrical vector modes. The correspondence between the linearly polarized modes and the conventional cylindrical vector modes is shown below (**Table 1**).

2.3.3 OAM modes

Optical fiber can support OAM modes by correctly superposing the even and odd modes for each $HE_{l,m}$ and $EH_{l,m}$ vector mode with $\pm(\pi/2)$ phase shift [27, 28]. Taking into consideration the circular polarization of OAM states (spin); OAM modes are denoted as $OAM_{\pm l,m}^{\pm}$ where $\pm$ superscript describes the spin angular momentum (circular polarization), l and m subscript denote the azimuthal and

Cylindrical vector modes	LP modes	Cylindrical vector modes	LP modes
HE_{1m}^{odd}	LP_{0m}^{y}	$HE_{2m}^{even} - TM_{0m}$	LP_{0m}^{eveny}
HE_{1m}^{even}	LP_{0m}^{x}	$HE_{(l+1)m}^{odd} + EH_{(l-1)m}^{odd}$	LP_{lm}^{oddx}
$HE_{2m}^{odd} + TE_{0m}$	LP_{1m}^{oddx}	$HE_{(l+1)m}^{odd} - EH_{(l-1)m}^{odd}$	LP_{lm}^{oddy}
$HE_{2m}^{odd} - TE_{0m}$	LP_{1m}^{oddy}	$HE_{(l+1)m}^{even} + EH_{(l-1)m}^{even}$	LP_{lm}^{evenx}
$HE_{2m}^{even} + TM_{0m}$	LP_{0m}^{evenx}	$HE_{(l+1)m}^{even} - EH_{(l-1)m}^{even}$	LP_{lm}^{eveny}

Table 1.
The correspondence between LP modes and the CV modes.

radial indices respectively. l is the topological number (number of twist in intensity profile), m describes the number of nulls radially (rings) in the intensity profile of the OAM mode. The magnitude of SAM equal $\pm s\hbar$ where $s = +1$ (left) or $s = -1$ (right). The magnitude of OAM equals $\pm l\hbar$. The total angular momentum AM is the sum of SAM and OAM with a magnitude of $(\pm l \pm s)\ \hbar$.

For the $TM_{0,m}$ and $TE_{0,m}$ modes, the combination between them with a $\pm$ ($\pi/2$) phase shift, carries the same magnitude of SAM and OAM but with opposite sign, making the total angular momentum equal to zero. This mode is not stable and cannot propagate, because the propagation constants of $TE_{0,m}$ and $TM_{0,m}$ modes are different. Therefore, we call this an unstable vortex. OAM modes made from $HE_{1,m}$ modes would have a spin, but no topological charge ($l = 0$). Therefore, this is not a true OAM mode, but simply a vector mode with circular polarization. However, we will consider it as $OAM_{0,m}$, in a more general definition.

OAM modes made from $HE_{l,m}$ modes are rotating in the same direction as the spin (aligned spin-orbit modes), and OAM modes made from $EH_{l,m}$ modes are rotating in the opposite direction as the spin (anti-aligned spin-orbit modes). If we take an even and an odd mode, with a $\pi/2$ phase difference, and we sum the fields (expressions 1.5 and 1.6), we can get as a resulting field:

$$\begin{cases} E_r = e_r(r)\, exp\,(\pm j\nu\phi) \\ E_\phi = je_\phi(r)\, exp\,(\pm j\nu\phi) \\ E_z = e_z(r)\, exp\,(\pm j\nu\phi) \end{cases} \tag{10}$$

The synthetic formula are as given in the following expressions

$$\begin{cases} HE^{even}_{l+1,m} \pm i \times HE^{odd}_{l+1,m} = OAM^{\mathrm{L/R}}_{\pm l,m} \\ EH^{even}_{l-1,m} \pm i \times EH^{odd}_{l-1,m} = OAM^{\mathrm{R/L}}_{\pm l,m} \\ \mathrm{TM_{0m}} \pm \mathrm{jTE_{0m}} = \mathrm{OAM^{\pm}_{\mp 1m}} \end{cases} \tag{11}$$

To summarize, for a given topological charge l, there are four possible *OAM* modes: two different spin rotation, and two different phase rotation. This is illustrated in **Figure 4**. The only exceptions are for $OAM_{\pm 1,m}$, where spin and topological charge always have the same sign, and for $OAM_{0,m}$, where there is no topological charge (only spin) [28].

Moreover, others OAM construction formulas are explored based on two spatially orthogonal linear polarized (LP) modes owning orthogonal polarization

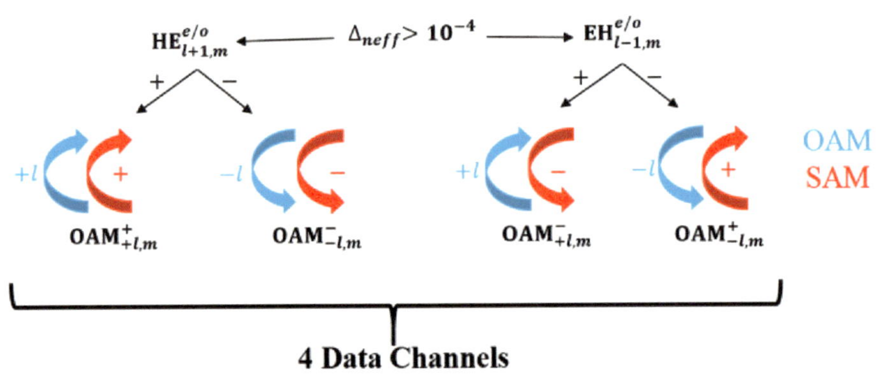

Figure 4.
The four OAM mode degeneracies (reproduced from [28]).

directions (with a $\pm\,\pi/2$ phase shift) which can be obtained by solving the scalar version of Maxwell equation (the scalar Helmholtz (wave) equation) under the weakly guiding approximation [29]. The LP-OAM synthetic formula are as follows:

$$\left\{\begin{matrix} LP_{lm}^{ax} \pm iLP_{lm}^{bx} \\ LP_{lm}^{ay} \pm iLP_{lm}^{by} \end{matrix}\right\} = F_{l,m}(r).\left\{\begin{matrix} \vec{x}OAM_{\pm l,m} \\ \vec{y}OAM_{\pm l,m} \end{matrix}\right\} \tag{12}$$

where $\vec{x}$ and $\vec{y}$ are the linear polarization along the x-axis and y-axis respectively, $F_{l,m}(r)$ is the radial field distribution. The difference between *OAM* modes generated from fiber vector modes (*CV-OAM*) possess circular polarization while those generated from LP modes (*LP-OAM*) are the linear polarization (has no *SAM*).

3. OAM-MDM through optical fibers

OAM has seen application in optical communication due to the theoretically unprecedented quantities of data that can be modulated, multiplexed, transmitted and demultiplexed through either free space link (FSO as Free Space Optics), or optical fibers. Optical communications has exploited the physical dimension of optical signal to encode and transmit individual/separate/independent data stream through the same transmission medium (optical fiber or FSO). Since, the OAM is linked to the spatial phase distribution of light beam, it has been included under the space dimension as a subset or embodiment of SDM (space division multiplexing). In addition, since OAM is independent of wavelength, quadrature, and polarization, it provides an additional dimension for encoding information [30, 31]. The interest on OAM in communication (including optical, radio, underwater) has grown dramatically. **Figure 5(a)** and **(b)**, which highlights the number of published papers (conferences paper, books, journal papers and patents), translates that huge interest. In **Figure 5(a)**, we plot the number of published papers dealing with OAM in optical communication in last decade while **Figure 5(b)** shows the number of papers dealing with OAM in optical fibers, both are according to *Google Scholar*.

The worldwide backbone of high-capacity wired communications is optical fiber. The uses of OAM basis in optical fiber was a challenge to communication community. For a long time, optical fibers were only used for the generation or the transformation of OAM modes, and not for supporting their transmission [32]. The notion of transmitting OAM modes was demonstrated (theoretically, numerically,

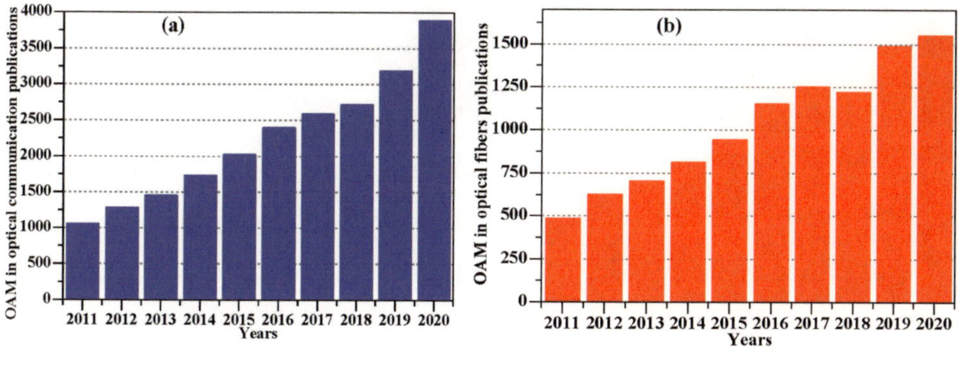

Figure 5.
Number of papers published dealing with (a) OAM in optical communication over ten years (ranging from 2011 to 2020), (b) OAM in optical fibers over the same period (according to Google scholar).

and experimentally) through conventional optical fibers (classical deployed fibers), or specialty fibers that have been specifically designed to transmit robust OAM modes. In the following, we present kinds of optical fibers based on the consideration of their refractive indexes, (e.g. graded, step, ring, etc.), geometrical features (MMF, SMF, and FMF etc.) and transmission caracteristics (MDM, CDM, PCF, kind of appropriate modes, etc.) and so on. We highlight the main design and principles results achievements.

3.1 Conventional fibers

Two examples of conventional optical fibers are multimode fiber MMF (e.g. OM1, OM2, OM3, OM4) where generally their refractive index are graded (GIF) and single mode fiber (e.g.G_{652}) where the profile is step index (SMF). Conventional MMFs have large cores that are usually approximately 50 μm and can support hundreds of modes. Due to severe inter-modal dispersion limitations, MMF were replaced by single mode fibers (SMFs) that have a relatively small core radius (not exceeding 10 μm). The refractive indexes of both fibers (OM3, and G_{652} defined by ITU-T) are depicted in **Figure 6(a)** and **(b)**.

The most commonly used modal basis for fibers are LP modes. LP modes are not exact fiber modes, and can be simply viewed as combinations of fiber eigenmodes transverse (TE, TM, HE and EH) as indicated above.

Other type of fibers are few mode fibers (FMF) which consist of an improved version of MMF. They support a limited number of modes, as one of the key components for SDM for optical networks. The first paper that mentioned the possibility of transmitting OAM modes through optical fiber is from Alexeyev et al. in 1998 [33]. The authors demonstrated that the solution for OAM modes could exist in optical fibers (MMF). Considering the propagation of OAM modes through the cited fibers, the analysis of OAM in conventional graded index multimode fiber was reported (theoretically and numerically) [34]. In that paper, Chen and his co-authors presented a comprehensive analysis of the ten-OAM modes groups supported in OM3, including mode coupling, chromatic dispersion, differential group delay, effective mode area and nonlinearities.

Later on, the same team demonstrated experimentally the transmission of four-OAM mode group in OM3 MMF using mode exciting and filtering elements at the 2-fiber extremity. Moreover, they demonstrated two OAM mode groups transmission over 2.6-km MMF with low crosstalk free of MIMO-DSP [35]. In 2018, Wang et al. reported the successful transmission of OAM modes over 8.8-km OM4 MMF [36]. Wang and co-workers demonstrates a 120-Gbit/s quadrature phase-shift

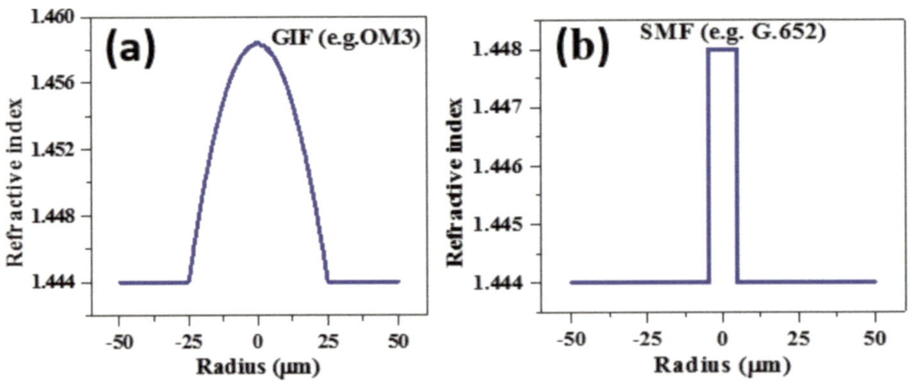

Figure 6.
Refractive indexes of (a) graded index fiber (multimode fiber) and (b) step index fiber (single mode fiber).

keying (QPSK) signal transmission over 8.8-km OM 4 MMF with 2×2 and 4×4 MIMO-DSP. In second stage, they demonstrate the data-carrying two OAM mode groups (6 OAM states) multiplexing transmission over 8.8 km MMF without MIMO equalization.

The OAM in SMF (ITU-T G.652) was investigated, in [37]. The investigation was performed over 3 visible wavelengths (red at 632.8 nm, green at 532 nm, and blue at 476.5 nm) when G.652 becomes a few mode fiber. The synthetized OAM modes was investigated through effective mode area, nonlinearity, tolerance to fiber ellipticity and bending. The authors analyzed and estimated the fiber attenuation and bandwidth/capacity for OAM modes over six levels of wavelengths.

Few mode fibers (FMFs) with classical refractive index profile (step/graded), was used to transmit OAM modes. The transmission of OAM modes over FMF required a MIMO-DSP in combination with coherent detection to equalize the intermodal crosstalk. It was demonstrated in [38] the transmission of four OAM beams over 5-km FMF. Each transmitted OAM state carrying 20 Gbits/s QPSK data. MIMO DSP was used to mitigate the mode coupling effects. A graded index few mode fiber has been designed in [39] in order to support 10 OAM orders with high purity ($\geq 99.9\%$) enabling low intermodal crosstalk ($\leq$-30 dB). Later, in [40] Wang et al. demonstrated the viability of OAM modes transmission over both 50-km and 10-km FMFs. By adopting LDPC codes, the DMD and mode coupling was improved. In [41], Zhu et al. proposed and demonstrated a heterogeneous OAM based fiber by splicing 2 FMFs and MMF (OM3). Over 2 OAM modes, Zhu and co-workers transmit 20-Gbit/s QPSK data without MIMO-DSP. Recently, we proposed a family of graded index few mode fibers (four fibers) that supports 12 OAM states [42]. The evaluated differential group delay (DGD) and OAM purity demonstrate the viability of proposed fibers for short/medium haul connections.

3.2 OAM specialty fibers

OAM has changed the common features of optical fiber design guidelines. Cutting with the often-classical notion for imposing the center core to be the highest index of refractive (graded & step). In addition, the improvement of optical fibers fabrication technologies (materials & schemes) has made the fibers characterization no more challenging. New optical fibers with complicated shapes and high refractive index contrast have been experimentally characterized (demonstrated). The Modified Chemical Vapor Deposition (MCVD) is in principle one of the most fiber fabrication method that has been extensively used.

3.2.1 OAM-fibers recommendations and design guidelines

Mainly three common features between OAM specialty fibers are identified. The first consists of the high contrast between core and cladding refractive indexes (jumps/contrasts) increasing the mode effective indices separation (Δn_{eff}), hence enabling low induced crosstalk. The same feature involves the formation of OAM modes from cylindrical vector modes and avoid them to couple into LP modes. It is proved that minimum Δn_{eff} of 10^{-4} is enough to keep robust OAM modes. This key value guarantees the minimum interaction between channels and prevents mode coupling inducing channels crosstalk XT. It has been demonstrated that through MCVD, a contrast of 0.14 is achievable with GeO_2-SiO_2 composition [43]. The second is about the refractive index profile that matches the donut shape of intensity profile of OAM mode (Ring shape: **Figure** 7). Thus, the Ring shaped (known also as depressed core fibers) has been extensively designed in OAM context instead of solid core fibers. Finally, the interfaces between fiber core and cladding preferred

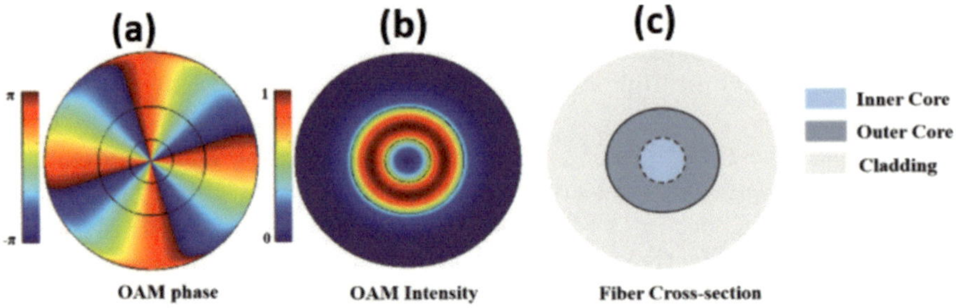

Figure 7.
OAM mode (a) phases pattern (e.g. $OAM_{4,1}$), (b) normalized intensity, (c) fiber cross-section with ring shape.

to be smoothed (instead of step (abrupt variation)) in order to eliminate the spin-orbit-coupling inducing OAM mode purity impairment and intrinsic crosstalk.

3.2.2 Vortex fibers (VFs)

The first specialty FMF designed for OAM modes is vortex fiber [44]. The designed fiber possess a good separation between co-propagating modes. Vortex fiber was first introduced to create cylindrical vector beams represented by $TE_{0,1}$ and $TM_{0,1}$ modes (also known as polarization vortices). Proposed by Ramachandran and al., Vortex fiber has a central core able to transmit the fundamental mode, surrounded by a lower trench, and an outer ring able to transmit the first OAM mode group. In first experience, they reported a transmission through more than 20 m fiber. Two years later, transmission of OAM through a 1 km fiber was reported [45–49]. **Figure 8** reported an optical microscope image of the end facet of the vortex fiber and the numerically calculated properties of the vortex fiber. All the experiments on OAM modes on the designed vortex fiber were summarized in [48].

3.2.3 Air Core fibers (ACFs)

Air core fiber (ACF) was proposed in [50]. 12 OAM modes were transmitted through 2 m of the fabricated ACF. Later on, 2 OAM modes were transmitted over 1 km of the fiber [51]. Among the main contributions, the authors demonstrated that OAM modes with higher l value are less sensitive to perturbations like bends and twists [52].

Within ROAM (revolution orbital angular momentum) project (EU H2020), Laval University (COPL) proposed and fabricated an ACF that achieved the record

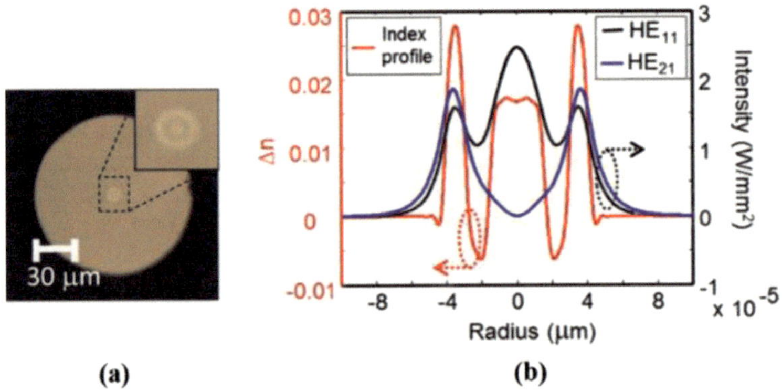

Figure 8.
(a) Optical microscope image of vortex fiber, (b) numerically calculated properties [48].

of OAM modes transmitted through an optical fiber. Benefiting from the high refractive index contrast (air/silica), the fabricated fiber supports the transmission of 36 OAM states [53]. **Figure 9** shows the refractive index of ACF. Nevertheless, the designed fiber possess a very high loss (up to few dBs per meters) which make it unsuitable for communication. Recently, 10.56 Tbit/s has been demonstrated over 1.2 km ACF, without MIMO DSP, by carrying data over 12 OAM modes combined with wavelength division multiplexing (WDM) [54]. Latest air core ring fiber is designed to support more than 1000 OAM modes (using As_2S_3 as ring material) across wide wavelength band covering S, O, E, S, C, and L Bands [55].

3.2.4 Inverse parabolic graded index fibers (IPGIFs)

Ung et al. proposed the inverse parabolic graded index FMF (IPGIF) to support OAM modes [56, 57]. The refractive index of IPGIF is given by the following expression:

$$n(r) = \begin{cases} n_2\sqrt{(1 - 2N\Delta\left(\frac{r^2}{a^2}\right)} & 0 \le r \le a \quad (core) \\ n_3 & r \ge a \quad (cladding) \end{cases} \tag{13}$$

Where a, n_1, n_2, n_3 are the core radius, the refractive index at the core cladding interfaces, the refractive index at the core center and the refractive index of the cladding, respectively. The parameter N controls the shape of the IPGF. The refractive index of IPGF is presented in **Figure 10**. Based on a first-order perturbations, the authors highlighted the factors (refractive index, core radius and curvature shape) that directly related to enhance the intermodal separation in proposed IPGIF. Large refractive index gradient, high transverse field amplitude and large field variation are reasons of high intermodal separation enabling low crosstalk.

The designed IPG-FMF possess a good effective indices separation ($\Delta n_{eff} > 2.1 \times 10^{-4}$) between its supported vector modes, and the transmission of eight OAM states ($OAM_{\pm0,1}$, $OAM_{\pm1,1}$ and $OAM_{\pm2,1}$) was demonstrated over *1 m* which makes IPGIF suitable for short distances MDM transmission. On the other hand, the transmission of $OAM_{\pm1,1}$ over more than 1 km was demonstrated by experiment, which makes the novel fiber as a promising candidate for long-distance

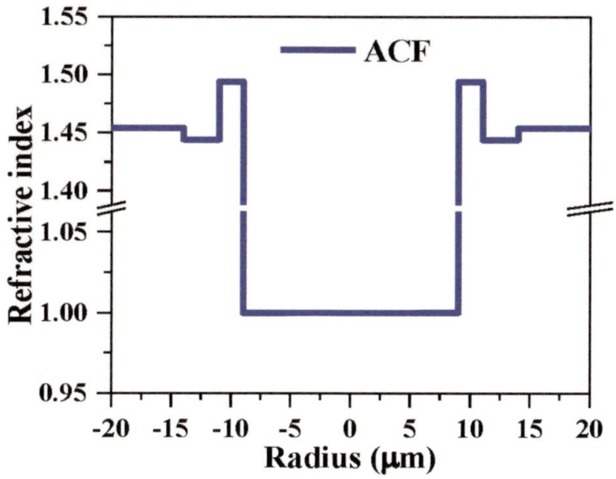

Figure 9.
Refractive index profile of an air (hollow) core fiber (ACF).

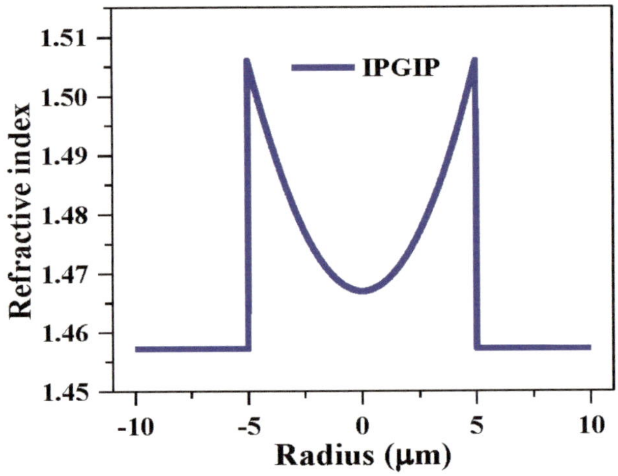

Figure 10.
Refractive index profile of inverse parabolic graded index fiber (IPGIF).

OAM based MDM multiplexing system. Later on (2017), the multiplexing/transmission and demultiplexing of 3.36 Tbits/s was demonstrated over 10-meters inverse parabolic graded index fiber by using four OAM modes and 15 wavelengths (WDM) [58].

3.2.5 Ring Core fibers (RCFs)

Due to the emerging interest in OAM-guiding fibers, already designed fiber for LP modes was investigated through OAM. The Ring core fiber (RCF), which has been introduced to minimize the differential group delay between LP modes, was tailored to support and transmit OAM modes. This interest on ring core fiber come from its refractive index profile that closely matches the annular intensity profile of OAM beams (**Figure 11**). C. Brunet et al. present an analytic tool to solve the vector version of Maxwell equations in RCF [59]. A fully vectorial description was reported in order to better tailor the RCF to OAM context. Using the modal map developed in [59], the group designed and manufactured a family of RCF (five fibers) suitable for OAM transmission [60]. In [61] S. Ramachandran et al. demonstrated the stability of OAM modes in RCF.

Recently, an RCF supported 50 OAM states divided into 13 mode groups (MGs) has been numerically investigated using small MIMO DSP blocks [62]. Experimentally, the transmission of two OAM mode-group is demonstrated over a 50 km ring core fiber without the use of MIMO DSP [63]. Emerging papers considering the

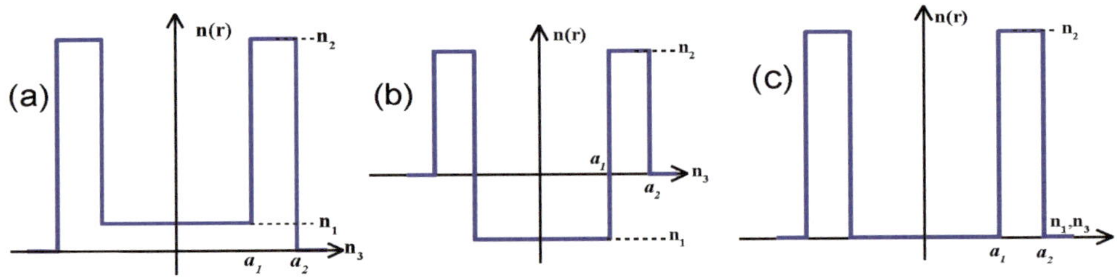

Figure 11.
Examples of RCF refractive index profiles: (a) RCF (higher center) (b) RCF (lower center) and (c) RCF, a_1 *and* a_2 *are inner and outer core radius respectively.*

design of RCFs and the propagation demonstration of OAM modes through it that we should mentions [64, 65].

3.2.6 Graded index ring Core fibers (GI-RCFs)

The ring notion touched the graded shape and a family of graded index ring core fibers (GI-RCF) has been proposed, designed and fabricated to support OAM mode group. **Figure 12** shows the refractive index of GI-RCF. In [66], Zhu and co-workers designed and fabricated the GI-RCF for OAM modes. The fiber supports 22 OAM modes with low insertion loss (less than 1 dB/km). The crosstalk between the highest order mode groups is less than 14 dB after10-km propagation. With such fiber, a successful transmission of 32 Gbaud QPSk-data overall 80 channels is experimentally demonstrated. A transmission capacity of 5.12 Tbits/s and a spectral efficiency of 9 bit/s/Hz, over 10 km propagation was reported [67].

The second demonstration was performed over 18-km propagation. Recently, the same group demonstrate the transmission of 12 Gbaud (8QAM) over 224 channels (2 OAM $\times$ 112 wavelengths). A transmission capacity of 8.4 Tbits/s was achieved without MIMO DSP because of the large high-order mode group separation of the OAM fiber [68].

To increase even further the capacity of the fiber link, OAM transmission was reported over uncoupled multi core fibers. While a complete review on this topic exceed the scope of this chapter, we can nevertheless mention some contributions. Li and Wang designed seven-ring core fiber (MOMRF) supporting 154 data-channels in total (22 modes$\times$ 7 rings) [69]. The proposed fiber featuring low-level inter-ring crosstalk (−30 dB for a 100-km-long fiber) and intermodal crosstalk over a wide wavelength range (1520–1580 nm). Later, in [70], Li and Wang proposed a compact trench multi OAM ring fiber (TA-MOMRF) with 19 rings each supporting 22 modes (18 OAM states). The authors stated that such fiber is suitable for long distance OAM transmission enabling Pbit/s total transmission capacity and hundreds bit/s/Hz spectral efficiency. In [71], the authors proposed a coupled multi core fibers to support OAM modes (multi-orbital-angular-momentum (OAM)

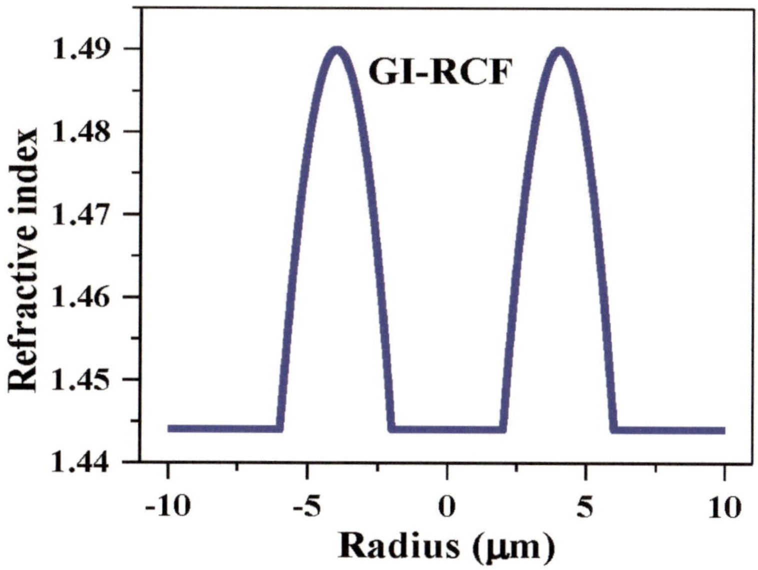

Figure 12.
Design of the GIRCF.

multicore supermode fiber (MOMCSF). The designed supermode fiber show favorable performance of low mode coupling, low nonlinearity, and low modal dependent loss.

3.2.7 Inverse raised cosine few mode fibers (IRC-FMFs)

Using IPGI fiber as a benchmark, we proposed a novel fiber that is based on inverse raised cosine function (IRCF). The standard raised cosine function (RCF) when applied to a wideband signal steeply removes the high out-of-band signals, making the filtered signal highly purified. Moreover, RCF is used in the same context because it eliminates intersymbol interference [72]. The IRCF profile is given by the following expression [73]:

$$n(r)=\begin{cases} n_2 \quad if \;\; 0\le r\le a\dfrac{1-\alpha}{2} \quad (Core) \\ -\dfrac{1}{2(\mathrm{n}_2-\mathrm{n}_3)}\left(1+\cos\left[\left(\dfrac{\pi}{a\times\alpha}\right)\left(\mathrm{r}-a\dfrac{1-\alpha}{2}\right)\right]\right) \quad if \;\; a\dfrac{1-\alpha}{2}\le r\le a\dfrac{1+\alpha}{2} \quad (Core) \\ n_3 \quad if \;\; r\ge a\dfrac{1+\alpha}{2} \quad (Cladding) \end{cases} \tag{14}$$

where a is the core radius, n_1 and n_2 are respectively the maximum and the minimum refractive indices of the core, n_3 is the refractive index of the cladding ($\mathrm{r} > a$), and α is the profile shape. The refractive index of IRCF is shown in the **Figure 13**. The IRC profile is practically thinner (or more concentrated around the fiber axis) than the IPGI profile [73]. However, it is worthy to note that our profile becomes much smoother when reaching the maximum index value n_1. When compared with IPGI-FMF, the inverse-raised-cosine function offers a large modal separation. The enhanced separation is likely to hinder mode coupling, reducing the system crosstalk and improving the transmission. Moreover, IRC-FMF has the potential to handle OAM modes with high purity hence low intrinsic crosstalk [73, 74].

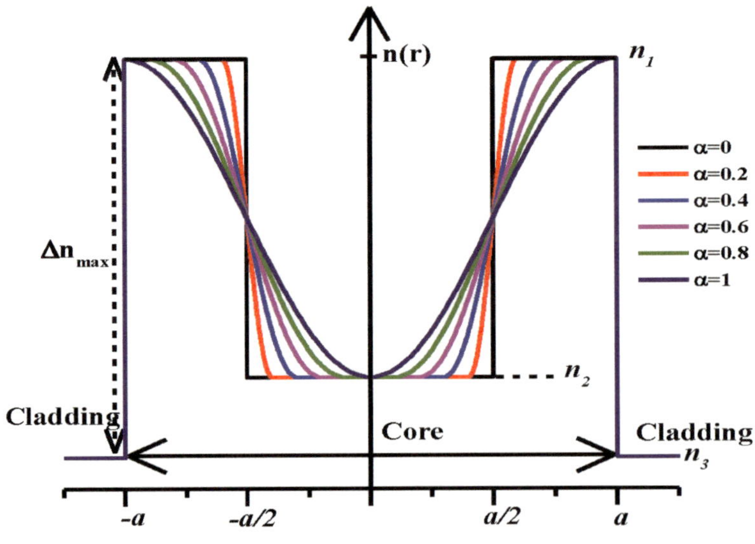

Figure 13.
Index profiles of the IRC fiber (solid lines), with α ranging from 0 to 1 (reproduced from 73).

3.2.8 Hyperbolic tangent few mode fibers (HTAN-FMFs)

Based on hyperbolic tangent function (HTAN), we proposed and designed a ring core few mode fiber that we refer to as hyperbolic tangent few mode fiber (HTAN-FMF). The function HTAN was not common in optical fiber profiling. It is widely used in various fields/domains such as digital neural networks, image processing, digital filters, and decoding algorithms [75–77] but not common in waveguide and optical fiber designs. Intuitively, one of the most attractive criteria in hyperbolic tangent function, used as an activation function in neural network, is its strong gradient centered around the inflection point (switch point). This is the same criteria required from an optical fiber profile in order to enhance the intermodal separation. The refractive index of HTAN-FMF is given by the following expression [78]:

$$n(r) = \begin{cases} n_2 \ if \ 0 \le |r| \le a\dfrac{(1-\alpha)}{2} \ (Core) \\ \dfrac{n_1+n_2}{2} + \dfrac{\Delta n}{2.Tanh(\pi)} \times \left[Tanh\left(\dfrac{\pi \times (r-a_1)}{a_1 \times \alpha}\right)\right] \ if \ a\dfrac{(1-\alpha)}{2} \le |r| \le a\dfrac{(1+\alpha)}{2} \ (Core) \\ n_1 \ if \ a\dfrac{(1+\alpha)}{2} \le |r| \le a \ (Core) \\ n_3 \ if |r| \ge a \ (Cladding) \end{cases} \quad (15)$$

Where n_1, n_2, n_3 are the refractive index at the core-cladding interface, at the core center, and at the cladding region, respectively. a, a_1 and α are the core radius, the half of core radius (a_1 = a/2) and the shape parameter respectively. Δn is the actual refractive index difference (i.e. $\Delta n = n_1 - n_2$) which corresponds to the extent of hyperbolic tangent function inside the core. The shape parameter α controls the shape behavior of HTAN function. The refractive index of HTAN is illustrated in **Figure 14**. The proposed HTAN-FMF achieves a wide intermodal separation (between cylindrical vector modes) especially between $TE_{0,1}$, $HE_{2,1}$, and $TM_{0,1}$

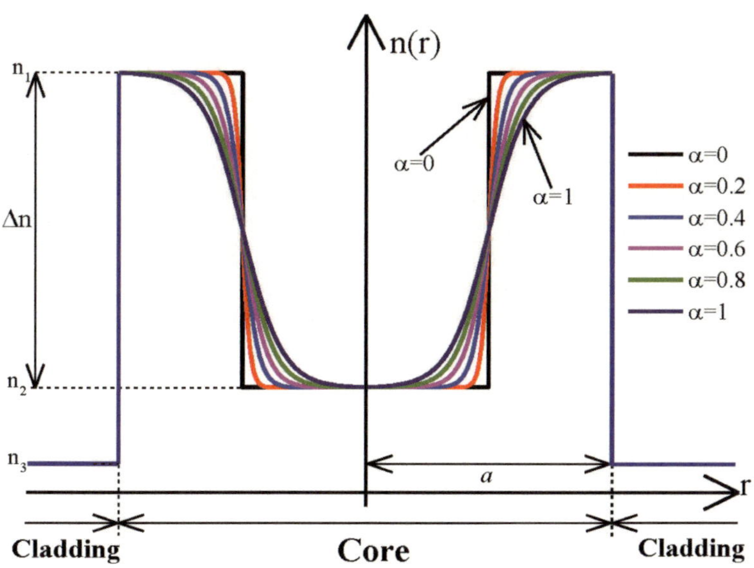

Figure 14.
Refractive index profile of HTAN fiber for different values of profile shape α [78].

(≥3 × 10^{-4}). This enables low-level crosstalk channels carrying data during propagation and outperforms what is existing in the literature [78]. On the other hand, even with an exterior abrupt variation, the inner smooth behavior of HTAN-FMF guarantees the enhancement of the obtained OAM mode purities (≥ 99.9%) leading to intrinsic crosstalk as minimum as −30 dB during propagation. Moreover, the obtained results in term of chromatic dispersion (max CD = −60 ps/(km.nm)), differential group delay (max DGD = 55 ps/m), and bending insensitivity, demonstrate that the HTAN-FMF could be a viable candidate for enhancing the transmission capacity and the spectral efficiency in next generation OAM mode division Multiplexing (OAM-MDM) systems [78].

3.3 Photonic crystal fibers

Photonic crystal fibers (PCF) has shown its design flexibility to guide appropriate OAM modes. With adjustable parameters, PCF can offer more flexible design structures to provide unique fiber properties. Due to that, several kinds of OAM-PCF with various structures (hexagonal, circular, kagome...) and materials (As_2S_3, SiO_2, polymer ...), having promising features have been designed and even fabricated. PCF have been proposed and fabricated to ensure good transmission quality of OAM modes. While a review on this topic exceed the scope of this thesis, we can nevertheless mention some details and contributions. PCFs supporting one, 2, 10, 12, 14, 26, 34, 42 and 48, first order OAM modes have been proposed featuring good transmission properties [79–89].

The race is still ongoing to increase the number of OAM modes in PCF featuring good transmission proprieties. To the best of our knowledge, the most supporting OAM modes number in a circular PCF reaches 110 over C + L communication bands [90]. The designed fiber featured large effective indices separation (are at the order of 10^{-3}), low nonlinear coefficient, low confinement loss (under 10^{-7} dB/m), and relatively flat chromatic dispersion. Such fiber could find potential application in high capacity OAM-MDM system. By analysis of these recent mosaic OAM-PCFs literature, we can come to the general requirements in PCF design that ensure good transmission quality of OAM modes in the following five points or guidelines [91–94].

- Fiber index profile that matches the intensity profile of OAM modes (ring shape).
- The supported modes belonging to the same OAM mode family should possess a large index separation (≥10^{-4}) to be free from complex and heavy multiple input multiple output digital signal processing (MIMO DSP) at the receiver side. This is achievable with high material contrast between the fiber core and the cladding. Instead of using pure SiO_2 as a background material for the PCF-fiber, other available materials could be used such as Silicon (Si), As_2S_3, and Polymer.
- Large core thickness is required targeting to increase the supported OAM mode number.
- The excited OAM modes should be of the first order. Hence, it is preferable to avoid exciting the higher radial orders modes because it causes trouble in multiplexing and demultiplexing operations due to the intensity and phase variety distribution.

- The guided OAM modes would possess good transmission features such as low confinement loss, flat dispersion, large effective mode area, and low nonlinear coefficient over a large wavelength range (at least covering C + L bands defined by ITU-T).

4. Conclusion

In this chapter, we have attempted to provide recent advances in SDM based Optical fibers. We showed that SDM is currently the unexhausted technology that can deal with the capacity need and boost data traffic. Furthermore, an interesting embodiment of SDM, which is based on carrying data on fiber modes (MDM) has been presented and discussed. The different mode basis supported in optical fibers are presented and discussed. Furthermore, we reviewed the research activities that are based on harnessing OAM modes to encode data channels either in classical optical fibers (i.e. with classical refractive index profiles) or in special fibers with appropriate ring profiles. We presented the main research activities and recent trends in OAM-MDM over the last ten years.

Author details

Alaaeddine Rjeb[1*], Habib Fathallah[2] and Mohsen Machhout[1]

1 Laboratory of Electronic and Microelectronic, Faculty of Sciences of Monastir, Physics Department, University of Monastir, Monastir, Tunisia

2 Laboratory of Artificial Intelligence and Data Engineering Applications, Faculty of Sciences of Bizerte, Computer Department, University of Carthage, Bizerte, Tunisia

*Address all correspondence to: alaaeddine.rjeb@gmail.com

References

[1] Cisco VNI Global IP traffic forecast "2016–2021".

[2] A. Chralyvy, Plenary paper: The coming capacity crunch. In: 2009 35th European Conference on Optical Communication, 2009, pp. 1–1.

[3] Ellis, A. D., Mac Suibhne, N., Saad, D., & Payne, D. N. Communication networks beyond the capacity crunch. (2016).

[4] R. Essiambre and R. Tkach. Capacity trends and limits of optical communication networks. In Proceedings of the IEEE, May 2012, vol. 100, no. 5, pp. 1035–1055.

[5] Richardson, D. J., Fini, J. M., & Nelson, L. E. Space-division multiplexing in optical fibres. Nature photonics. 2013, 7(5), 354–362.

[6] Saridis, G. M., Alexandropoulos, D., Zervas, G., & Simeonidou, D. Survey and evaluation of space division multiplexing: From technologies to optical networks. IEEE Communications Surveys & Tutorials. 2015; 17(4), 2136–2156.

[7] Saitoh, K., & Matsuo, S. Multicore fiber technology. Journal of Lightwave Technology. 2016; 34(1), 55–66.

[8] Yaman, F., Bai, N., Zhu, B., Wang, T., & Li, G. Long distance transmission in few-mode fibers. Optics Express. (2010); 18(12), 13250–13257.

[9] L.A. Rusch, M. Rad, K. Allahverdyan, I. Fazal, E. Bernier, Carrying data on the orbital angular momentum of light, IEEE Commun. Mag. (2018), 56 (2) 219–224.

[10] Bozinovic, N., Yue, Y., Ren, Y., Tur, M., Kristensen, P., Huang, H., & Ramachandran, S. Terabit-scale orbital angular momentum mode division multiplexing in fibers. Science. (2013), 340(6140), 1545–1548.

[11] Wang, J., Yang, J. Y., Fazal, I. M., Ahmed, N., Yan, Y., Huang, H., & Willner, A. E. Terabit free-space data transmission employing orbital angular momentum multiplexing. Nature photonics, (2012), 6(7), 488–496.

[12] "Sumitomo Electric." [Online]. Available: http://www.sumitomoelectric.com/.

[13] "OFS." [Online]. Available: http://www.ofsoptics.com.

[14] "Samtec." [Online]. Available: http://www.samtec.com.

[15] T. Hayashi, T. Taru, O. Shimakawa, T. Sasaki, and E. Sasaoka, "Design and fabrication of ultra-low crosstalk and low-loss multi-core fiber," Opt. Express. 2011, vol. 19, no. 17, pp. 16576–16592.

[16] Kingsta, R. M., & Selvakumari, R. S. A review on coupled and uncoupled multicore fibers for future ultra-high capacity optical communication. Optik. (2019), 199, 163341.

[17] M. Koshiba, K. Saitoh, and Y. Kokubun, "Heterogeneous multi-core fibers: proposal and design principle," IEICE Electron. Express. 2009, vol. 6, no. 2, pp. 98–103, Jan.

[18] Y. Kokubun and M. Koshiba, "Novel multi-core fibers for mode division multiplexing: proposal and design principle," IEICE Electron. Express. Apr. 2009, vol. 6, no. 8, pp. 522–528.

[19] C. Xia, N. Bai, I. Ozdur, X. Zhou, and G. Li, "Supermodes for optical transmission," Opt. Express. Aug. 2011, vol. 19, no. 17, pp. 16 653–16 664.

[20] Zhu, B., Taunay, T. F., Fishteyn, M., Liu, X., Chandrasekhar, S., Yan, M. F., & Dimarcello, F. V. 112-Tb/s space-division multiplexed DWDM

transmission with 14-b/s/Hz aggregate spectral efficiency over a 76.8-km seven-core fiber. Optics Express. (2011), 19(17), 16665–16671.

[21] T. Hayashi, T. Taru, O. Shimakawa, T. Sasaki, and E. Sasaoka, Low-Crosstalk and Low-Loss Multi-Core Fiber Utilizing Fiber Bend. In: Optical Fiber Communication Conference/National Fiber Optic Engineers Conference 2011, OSA Technical Digest (CD) (Optical Society of America, 2011).

[22] H. Takara, A. Sano, T. Kobayashi, H. Kubota, H. Kawakami, A. Matsuura, Y. Miyamoto, Y. Abe, H. Ono, K. Shikama, Y. Goto, K. Tsujikawa, Y. Sasaki, I. Ishida, K. Takenaga, S. Matsuo, K. Saitoh, M. Koshiba, and T. Morioka. 1.01-Pb/s (12 SDM/222 WDM/456 Gb/s) crosstalk-managed transmission with 91.4-b/s/Hz aggregate spectral efficiency. In: the European Conf. Exhibition Optical Communication, Amsterdam, the Netherlands, 2012.

[23] K. Igarashi, et al., 1.03-Exabit/s· km Super-Nyquist-WDM Transmission over 7,326-km Seven-Core Fiber. ECOC-2013, PD1.E.3.

[24] S. Berdague, and P. Facq, Mode Division Multiplexing in Optical Fibers. App. Opt. (1982), 24(11), 1950–1955.

[25] Russell, P. Photonic crystal fibers. Science, (2003), 299(5605), 358–362.

[26] J. Bures, Guided Optics, ser. Physics textbook. Wiley, 2009.

[27] Zhang, H., Mao, B., Han, Y., Wang, Z., Yue, Y., & Liu, Y. Generation of orbital angular momentum modes using fiber systems. Applied Sciences, (2019), 9(5), 1033.

[28] Brunet, C., & Rusch, L. A. Optical fibers for the transmission of orbital angular momentum modes. Optical Fiber Technology, (2017), 35, 2–7.

[29] Ramachandran, S., & Kristensen, P. Optical vortices in fiber. Nanophotonics. (2013), 2(5–6), 455–474.

[30] Yao, A. M., & Padgett, M. J. Orbital angular momentum: origins, behavior and applications. Advances in Optics and Photonics. (2011), 3(2), 161–204.

[31] Padgett, M. J. Orbital angular momentum 25 years on. Optics express. (2017), 25(10), 11265–11274.

[32] R. Kumar, D. S. Mehta, A. Sachdeva, A. Garg, P. Senthilkumaran, and C. Shakher. Generation and detection of optical vortices using all fiber-optic system. Optics Communications. 2008, vol. 281, no. 13, pp. 3414–3420.

[33] A. N. Alexeyev, T. A. Fadeyeva, and A. V. Volyar, "Optical vortices and the flow of their angular momentum in a multimode fiber," Semiconductor Physics, Quantum Electronics & Optoelectronics. 1998, vol. 1, no. 1, pp. 82–89.

[34] Chen, S., & Wang, J. Theoretical analyses on orbital angular momentum modes in conventional graded-index multimode fibre. Scientific Reports. (2017), 7(1), 3990.

[35] Zhu, L., Wang, A., Chen, S., Liu, J., Mo, Q., Du, C., & Wang, J. Orbital angular momentum mode groups multiplexing transmission over 2.6-km conventional multi-mode fiber. Optics express. (2017), 25(21), 25637–25645.

[36] Wang, A., Zhu, L., Wang, L., Ai, J., Chen, S., & Wang, J. Directly using 8.8-km conventional multi-mode fiber for 6-mode orbital angular momentum multiplexing transmission. Optics express. (2018), 26(8), 10038–10047.

[37] Chen, S., & Wang, J. Characterization of red/green/blue orbital angular momentum modes in conventional G. 652 fiber. IEEE Journal

of Quantum Electronics. (2017), 53(4), 1–14.

[38] G. Milione, H. Huang, M. Lavery, A. Willner, R. Alfano, T. A. Nguyen, and M. Padgett, "Orbital-angular-momentum mode (de)multiplexer: a single optical element for MIMO-based and non-MIMO-based multimode fiber systems. In: Optical Fiber Communication Conference, OSA Technical Digest (online) (Optical Society of America, 2014).

[39] Zhang, Z., Gan, J., Heng, X., Wu, Y., Li, Q., Qian, Q., & Yang, Z. Optical fiber design with orbital angular momentum light purity higher than 99.9%. Optics express. (2015), 23(23), 29331–29341.

[40] Wang, A., Zhu, L., Chen, S., Du, C., Mo, Q., & Wang, J. Characterization of LDPC-coded orbital angular momentum modes transmission and multiplexing over a 50-km fiber. Optics express, (2016), 24(11), 11716–11726.

[41] Zhu, L., Wang, A., Chen, S., Liu, J., & Wang, J. Orbital angular momentum mode multiplexed transmission in heterogeneous few-mode and multi-mode fiber network. Optics letters. (2018), 43(8), 1894–1897.

[42] Rjeb, A., Seleem, H., Fathallah, H., & Machhout, M. Design of 12 OAM-Graded index few mode fibers for next generation short haul interconnect transmission. Optical Fiber Technology. (2020), 55, 102148.

[43] H. Doweidar. Considerations on the structure and physical properties of B2O3–SiO2 and GeO2–SiO2 glasses.J. Non-Cryst. Solids. (2011), 357(7), 1665–1670.

[44] S. Ramachandran, P. Kristensen, and M. F. Yan, Generation and propagation of radially polarized beams in optical fibers. Opt. Lett. Aug. 2009, vol. 34, no. 16, pp. 2525–2527,.

[45] N. Bozinovic, P. Kristensen, and S. Ramachandran, Long-range fiber-transmission of photons with orbital angular momentum. In: CLEO: 2011 - Laser Applications to Photonic Applications. Optical Society of America, 2011.

[46] Bozinovic, N., Kristensen, P., & Ramachandran, S. Are orbital angular momentum (OAM/vortex) states of light long-lived in fibers. In: Laser Science (p. LWL3). Optical Society of America.

[47] N. Bozinovic, S. Ramachandran, M. Brodsky, and P. Kristensen. Record-length transmission of entangled photons with orbital angular momentum (vortices). In: Frontiers in Optics, (2011, October). Optical Society of America, 2011.

[48] N. Bozinovic, Y. Yue, Y. Ren, M. Tur, P. Kristensen, H. Huang, A. E. Willner, and S. Ramachandran, Terabit-scale orbital angular momentum mode division multiplexing in fibers. Science. 2013, vol. 340, no. 6140, pp. 1545–1548.

[49] S. Golowich, P. Kristensen, N. Bozinovic, P. Gregg, and S. Ramachandran. Fibers supporting orbital angular momentum states for information capacity scaling. In: Proc. of FIO. OSA, 2012.

[50] P. Gregg, P. Kristensen, S. Golowich, J. Olsen, P. Steinvurzel, and S. Ramachandran, "Stable transmission of 12 OAM states in air-core fiber: In CLEO: 2013.Optical Society of America, 2013.

[51] P. Gregg, P. Kristensen, and S. Ramachandran. OAM stability in fiber due to angular momentum conservation. In CLEO: 2014. Optical Society of America, 2014.

[52] Gregg, P., Kristensen, P., & Ramachandran, S. Conservation of orbital angular momentum in air-core optical fibers. Optica. (2015), 2(3), 267–270.

[53] C. Brunet, P. Vaity, Y. Messaddeq, S. LaRochelle, and L. A. Rusch, Design, fabrication and validation of an OAM fiber supporting 36 states. Optics Express. (2014), 22, no 21, pp. 26117–26127.

[54] K. Ingerslev, P. Gregg, M. Galili, F. Da Ros, H. Hu, F. Bao, M. A. U. Castaneda, P. Kristensen, A. Rubano, L. Marrucci, and K. Rottwitt.12 mode, WDM, MIMO-free orbital angular momentum transmission. Opt. Exp. Aug. 2018, vol. 26, pp. 20225–20232.

[55] Y Wang, C. Bao, W. Geng, Y. Lu, Y. Fang, B. Mao, Y.-G. Liu, H. Huang, Y. Ren, and Z. Pan. Air-core ring fiber with >1000 radially fundamental OAM modes across O, E, S, C, and L bands. IEEE Access. 2020, vol. 8, pp. 68280–68287.

[56] Ung, P. Vaity, L. Wang, Y. Messaddeq, L. A. Rusch and S. LaRochelle. Few-mode fiber with inverse-parabolic graded-index profile for transmission of OAM-carrying modes. Optics Express. (July 2014), 22, no. 15, pp. 18044–18055.

[57] Ung, B., Wang, L., Brunet, C., Vaity, P., Jin, C., Rusch, L. A., & LaRochelle. S. Inverse-parabolic graded-index profile for transmission of cylindrical vector modes in optical fibers. In: Optical Fiber Communication Conference, (2014, March), (pp. Tu3K-4). Optical Society of America.

[58] J. Zhu, et al. 3.36-Tbit/s OAM and Wavelength Multiplexed Transmission over an Inverse-Parabolic Graded Index Fiber. In: CLEO 2017, San Jose, p. SW4I.3, May 2017.

[59] C. Brunet, B. Ung, P.-A. Bélanger, Y. Messaddeq, S. LaRochelle, and L. A. Rusch.Vector mode analysis of ring-core fibers: design tools for spatial division multiplexing. Journal of Lightwave Technology. (2014), 32.23, 4046–4057.

[60] C. Brunet, B. Ung, L. Wang, Y. Messaddeq, S. LaRochelle, and L. A. Rusch. Design of a family of ring-core fibres for OAM transmission studies. Optics Express. (2015), 23.8: 10553–10563.

[61] Ramachandran, S., Gregg, P., Kristensen, P., & Golowich, S. E. On the scalability of ring fiber designs for OAM multiplexing. Optics express. (2015), 23 (3), 3721–3730.

[62] S. Chen, S. Li, L. Fang, A. Wang, and J. Wang. OAM mode multiplexing in weakly guiding ring-core fiber with simplified MIMO-DSP. Opt. Exp. 2019, vol. 27, no. 26, pp. 38049–38060.

[63] R. Zhang, H. Tan, J. Zhang, L. Shen, J. Liu, Y. Liu, L. Zhang, and S. Yu. A novel ring-core fiber supporting MIMO-free 50 km transmission over high-order OAM modes. In: Proc. Opt. Fiber Commun. Conf. (OFC), 2019, pp. 1–3.

[64] Zhang, J., Wen, Y., Tan, H., Liu, J., Shen, L., Wang, M., & Yu, S. 80-Channel WDM-MDM transmission over 50-km ring-core fiber using a compact OAM DEMUX and modular 4× 4 MIMO equalization. In: Optical Fiber Communication Conference (pp. W3F-3). Optical Society of America, (2019, March).

[65] Banawan, M., Wang, L., LaRochelle, S., & Rusch, L. A. Quantifying the Coupling and Degeneracy of OAM Modes in High-Index-Contrast Ring Core Fiber. Journal of Lightwave Technology. (2020), 39(2), 600–611.

[66] Zhu, G., Chen, Y., Du, C., Zhang, Y., Liu, J., & Yu, S. A graded index ring-core fiber supporting 22 OAM states. In: Opto-Electronics and Communications Conference (OECC) and Photonics Global Conference (PGC), 2017 (pp. 1–3). IEEE.

[67] G. Zhu, et al. Scalable mode division multiplexed transmission over a 10-km ring-core fiber using high-order orbital

angular momentum modes. Optics Express. vol. 26, pp. 594–604.

[68] Zhu, L., Zhu, G., Wang, A., Wang, L., Ai, J., Chen, S. ... & Wang, J. 18 km low-crosstalk OAM+WDM transmission with 224 individual channels enabled by a ring-core fiber with large high-order mode group separation. Optics letters. (2018), 43(8), 1890–1893.

[69] Li, S., & Wang, J. Multi-orbital-angular-momentum multi-ring fiber for high-density space-division multiplexing. IEEE Photonics Journal. (2013), 5(5), 7101007–7101007.

[70] Li, S., & Wang, J. A compact trench-assisted multi-orbital-angular-momentum multi-ring fiber for ultrahigh-density space-division multiplexing (19 rings× 22 modes). Scientific reports. (2014), 4, 3853.

[71] S. Li and J. Wang. Supermode fiber for orbital angular momentum (OAM) transmission. Opt. Express. Jul. 2015, vol. 23, pp. 18736–18745.

[72] Proakis, J. G., Salehi, M., Zhou, N., & Li, X. (1994). Communication systems engineering (Vol. 2). New Jersey: Prentice Hall.

[73] Rjeb, A., Guerra, G., Issa, K., Fathallah, H., Chebaane, S., Machhout, M., & Galtarossa. A. Inverse-raised-cosine fibers for next-generation orbital angular momentum systems. Optics Communications. (2020), 458, 124736.

[74] Rjeb, A., Fathallah, H., & Machhout, M. Orbital Angular Momentum Mode Coupling Analysis due to Ellipticity and Birefringence in Inverse-raised Cosine Fiber. In: 2020 17th International Multi-Conference on Systems, Signals & Devices (SSD) (pp. 929–932). IEEE, (2020, July).

[75] Garg, G., Sharma, P. An Analysis of Contrast Enhancement using Activation Functions. International Journal of Hybrid Information Technology. (2014), 7(5), 2.

[76] Johansen, H. K., Sørensen, K., "Fast hankel transforms", Geophysical Prospecting, 27(4), 876–901, (1979).

[77] Goyal, A., Kwon, H. M. Hyperbolic tangent function avoided for encoded pilot low density parity check decoding. In: The European Conference on Wireless Technology, (pp. 149–152). IEEE, (2005, October).

[78] Rjeb, A., Fathallah, H., Khaled, I., Machhout, M., & Alshebeili, S. A. A Novel Hyperbolic Tangent Profile for Optical Fiber for Next Generation OAM-MDM Systems. IEEE Access. (2020), 8, 226737–226753.

[79] Wong, G. K. L., Kang, M. S., Lee, H. W., Biancalana, F., Conti, C., Weiss, T., & Russell, P. S. J. Excitation of orbital angular momentum resonances in helically twisted photonic crystal fiber. Science. (2012), 337(6093), 446–449.

[80] Yue, Y., Zhang, L., Yan, Y., Ahmed, N., Yang, J. Y., Huang, H., ... & Willner, A. E. (2012). Octave-spanning supercontinuum generation of vortices in an As 2 S 3 ring photonic crystal fiber. Optics letters, 37(11), 1889–1891.

[81] Zhang, H.; Zhang, W.; Xi, L.; Tang, X.; Tian, W.; Zhang, X. In: Proceedings of the Asia Communications and Photonics Conference, 2015, Hong Kong, China, pp. 1–3, 19–23 (November 2015).

[82] Zhang, H., Zhang, X., Li, H., Deng, Y., Zhang, X., Xi, L., ... & Zhang, W. A design strategy of the circular photonic crystal fiber supporting good quality orbital angular momentum mode transmission. Optics Communications. (2017), 397, 59–66.

[83] Zhang, H., Zhang, W., Xi, L., Tang, X., Zhang, X., & Zhang, X. A new type circular photonic crystal fiber for orbital angular momentum mode transmission.

IEEE Photonics Technology Letters. (2016), 28(13), 1426–1429.

[84] Hu, Z. A., Huang, Y. Q., Luo, A. P., Cui, H., Luo, Z. C., & Xu, W. C. Photonic crystal fiber for supporting 26 orbital angular momentum modes. Optics express. (2016), 24(15), 17285–17291.

[85] Tian, W., Zhang, H., Zhang, X., Xi, L., Zhang, W., & Tang, X. A circular photonic crystal fiber supporting 26 OAM modes. Optical Fiber Technology. (2016), 30, 184–189.

[86] Chen, C., Zhou, G., Zhou, G., Xu, M., Hou, Z., Xia, C., & Yuan, J. A multi-orbital-angular-momentum multi-ring micro-structured fiber with ultra-high-density and low-level crosstalk. Optics Communications. (2016), 368, 27–33.

[87] Xi, X. M., Wong, G. K. L., Frosz, M. H., Babic, F., Ahmed, G., Jiang, X., ... & Russell, P. S. J. Orbital-angular-momentum-preserving helical Bloch modes in twisted photonic crystal fiber. Optica. (2014), 1(3), 165–169.

[88] Zhang, H., Zhang, X., Li, H., Deng, Y., Zhang, X., Xi, L., ... & Zhang, W. A design strategy of the circular photonic crystal fiber supporting good quality orbital angular momentum mode transmission. Optics Communications. (2017), 397, 59–66.

[89] Rjeb, A., Habib Fathallah, Saleh Chebaane, Mohsen Machhout Design of Novel Circular Lattice Photonic Crystal Fiber suitable for transporting 48 OAM modes. Accepted for publication in Optoelectronics Letters (2021). [In Press].

[90] Zhang, L., Zhang, K., Peng, J., Deng, J., Yang, Y., & Ma, J. Circular photonic crystal fiber supporting 110 OAM modes. Optics Communications. (2018), 429, 189–193.

[91] Li, H., Ren, G., Zhu, B., Gao, Y., Yin, B., Wang, J., & Jian, S. Guiding terahertz orbital angular momentum beams in multimode Kagome hollow-core fibers. Optics letters. (2017), 42(2), 179–182.

[92] Zhang, H., Zhang, X., Li, H., Deng, Y., Xi, L., Tang, X., & Zhang, W. The orbital angular momentum modes supporting fibers based on the photonic crystal fiber structure. Crystals. (2017), 7(10), 286.

[93] Zhang, H., Zhang, X., Li, H., Deng, Y., Zhang, X., Xi, L., & Zhang, W. A design strategy of the circular photonic crystal fiber supporting good quality orbital angular momentum mode transmission. Optics Communications. (2017), 397, 59–66.

[94] Li, H., Zhang, H., Zhang, X., Zhang, Z., Xi, L., Tang, X. ... & Zhang, X. Design tool for circular photonic crystal fibers supporting orbital angular momentum modes. Applied optics. (2018), 57(10), 2474–2481.

Section 3

Optical Fiber Integration and Photonic Devices

Chapter 6

Optical Fiber Tweezers for the Assembly of Living Photonic Probes

Xing Li and Hongbao Xin

Abstract

Optical fiber tweezers, as a versatile tool for optical trapping and manipulation, have attracted much attention in cell trapping, manipulation, and detection. Particularly, assembly of living cells using optical fiber tweezes has become a significant attention. Advanced achievements have been made on the assembly of fully biocompatible photonic probes with biological cells, enabling optical detection in biological environment in a highly compatible manner. Therefore, in this chapter, we discuss the use of optical fiber tweezers for assembly of living photonic probes. Living photonic probes can be assembled by the trapping and assembly of multiple cells using optical fiber tweezers. These photonic probes exhibit high biocompatibility and show great promise for the bio-applications in bio-microenvironments.

Keywords: Optical fiber tweezers, living photonic probes, optical trapping, optical manipulation, cell assembly

1. Introduction

The development of optical fiber tweezers (OFTs) makes it a versatile candidate for optical trapping and manipulation of targets ranging from different dielectric particles to biological cells and biomolecules [1–3]. This is because OFTs possess exceptional advantages in manipulation flexibility, due to the simple structure with only optical fibers. This simple structure also avoids the use of a high numerical-aperture objective which is necessary for the light focusing in conventional optical tweezers system [4, 5]. It is much easier to handle and manipulate the microscopic objects after trapped with OFTs [6, 7]. And it is much more suitable for practical use such as in trapping, levitating and rotating of microscopic particles in different environments [8–10]. The OFTs tip can be inserted into thick samples and turbid media, which greatly increases the sample applicability. In addition, OFTs exhibit a low-cost manipulation technique and can also be integrated into small devices, such as optofluidic channels [11]. OFTs enable the trapping and manipulation of different single targets. For the further biological detection in bio-environments, it is highly desired to form biocompatible photonic probes that can minimize the physical damage to the biological samples. Unfortunately, most photonic probes are made from inorganic and artificial materials, which are incompatible and invasive when interfacing with biological systems. It is still a big challenge to find out a biomaterial to assemble biophotonic probes that are noninvasive and highly biocompatible to biological systems. Fortunately, it is found that living cells, which are abundant in the natural world, show the capability for light manipulation and propagation with high biocompatibility, and can thus be used for the assembly of living photonic probes. In this chapter, recent advances of OFTs in trapping and manipulating of cells, particularly in assembly of living photonic probes based on biological cells, were discussed. These formed living photonic probes provide a promising approach for bio-detection in biological environments with highly biocompatibility [12, 13].

2. Working principle of OFTs

OFTs, generally based on a tapered fiber probe, can be fabricated by drawing a commercial single-mode optical fiber through a flame-heating technique. The shape of OFTs tip can be controlled by controlling the heating temperature and the drawing speed. The operation principle of typical OFTs has been detailedly analyzed and described [14]. As schematically shown in **Figure 1a**, an OFT is immersed in water. D_A means the axial distance of a dispersed particle to the OFTs tip, while D_T means the transverse distance. With a laser beam launched into the OFTs, particle will be trapped and manipulated by the generated optical force. There two components of the optical force, *i.e.*, gradient force (F_g) and scattering force (F_s). F_g is directed to the region with stronger light intensity and

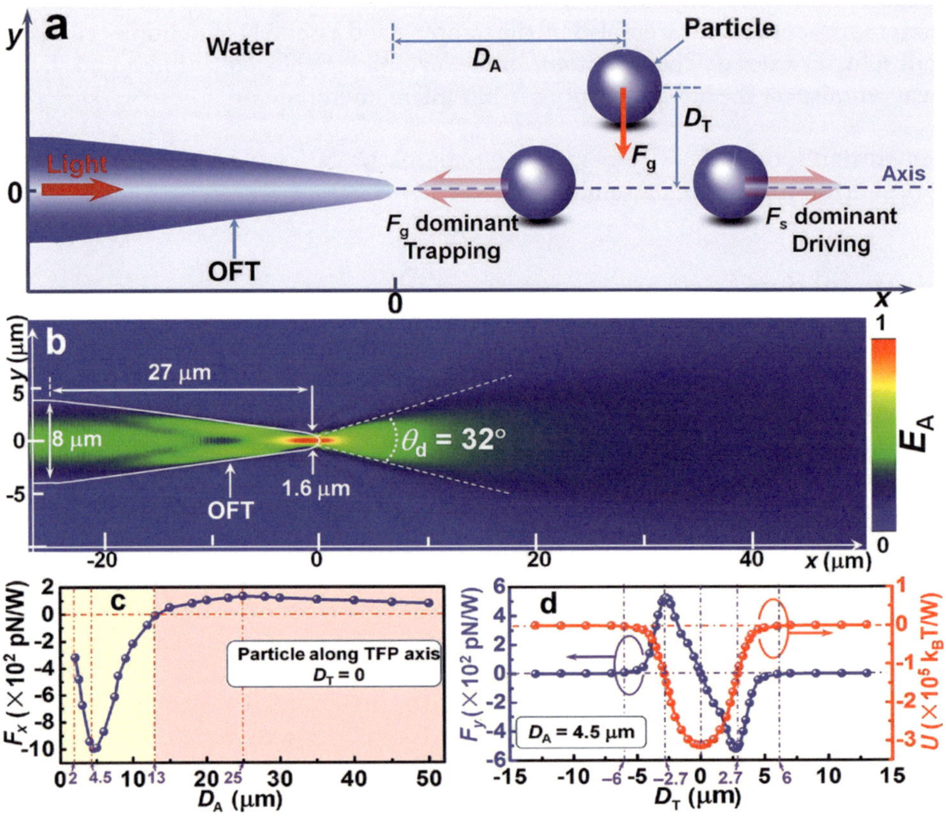

Figure 1.
Principle of a single optical fiber tweezers for trapping of particles [14]. (a) Schematic of particle manipulation by an OFT with light launched. (b) Simulated electric field amplitude (E_A) distribution by FDTD method. (c) Calculated optical force exerted on particles along the x direction. (d) Calculated optical force and trapping potential along the y direction.

is responsible to trap the particle, while F_s is directed along the light propagation and can push particles away from the OFTs tip. When a particle is near the axial axis of the OFTs, it will be trapped to the axis by F_g. For particle near the OFTs tip, the dominated F_g can trap the particle to the fiber tip. As the distance to the tip increases, F_s will become larger than F_g, and the dominated F_s will push the particle away from the fiber tip. The electric field amplitude (E_A) distribution around the OFTs was shown in **Figure 1b**, with a laser beam at a wavelength of 980 nm launched into the fiber probe. It can be seen that the light outputted from the OFTs is firstly focused at the tip and subsequently diverged out in water with a divergence angle of 32°. **Figure 1c** shows the calculated optical force exerted on a 3-μm silica particle along the x direction. It can be seen that, near the fiber tip, the force is negative, indicating a trapping force for particles. Therefore, particles near the fiber tip can be trapped by the OFTs. As the distance increases, the force is positive, indicating a driving force for particles. Therefore, particles can be pushed away by the OFTs. **Figure 1d** shows the calculated force and trapping potential in the y direction. It can be seen that the trapping potential on the axis is the smallest, and therefore particles beside the axis can be trapped at the axis. These optical forces enable the trapping capability of OFTs. By simply moving the fiber probe, the trapped particles can be manipulated in a highly flexible manner.

3. Manipulation of single cell and multiple cells by OFTs

OFTs can serve as a powerful tool for the trapping and manipulation of cells. Using *Escherichia coli* as an example, both single and multiple motile bacteria have been trapped and manipulated in a non-contact manner [15]. **Figure 2a** shows the experimental schematic for non-contact trapping of *E. coli* using OFTs. In this scenario, a laser beam at a wavelength of 980 nm was launched into the OFTs. A *E. coli* bacterium that was randomly swimming in the suspension was then trapped by the OFTs. The trapping was a non-contact trapping, and the bacterium was in the trapping position with several microns to the tip of the OFTs. During the trapping, the highly active bacterium was struggling around the trapping region. **Figure 2b–d** shows the detailed process for the trapping and struggling dynamics. The bacterium was trapped by the OFTs in a non-contact manner. However, due to the motility, the trapped bacterium was struggling after trapping. This phenomenon provides a new method for the studying of bacteria dynamics using OFTs.

In addition to the trapping and manipulation of single cells, OFTs can also be used for the trapping and assembly of multiple cells. For example, **Figure 3a** shows a schematic for the trapping and assembly of multiple *E. coli* cells in a microfluidic channel using OFTs [16]. Light output from the OFTs can trap the *E. coli* bacteria delivered by microfluidics. After a single bacterium was trapped, light can further propagate along the cell, and can be used for the trapping of other bacteria. Therefore, multiple bacteria can be trapped and assembled into cell chains with different lengths. To show the multiple trapping capability, **Figure 3b** shows the simulated light propagation along multiple cells. It can be seen that, light can propagate along the trapped cells, and the exerted optical force can be used for further trapping of other bacteria (**Figure 3c**). To experimentally demonstrate stable trapping and connecting of multiple *E. coli* cells with highly organized orientation, i.e., realization and retaining of *E. coli* cell–cell contact, the 980-nm wavelength laser with an optical power was launched into the fiber probe. **Figure 3d** shows the trapped multiple cells and formed cell chains with different numbers of cells at different input optical powers. By moving the fiber probe, the assembled cell chains can further be flexibly manipulated.

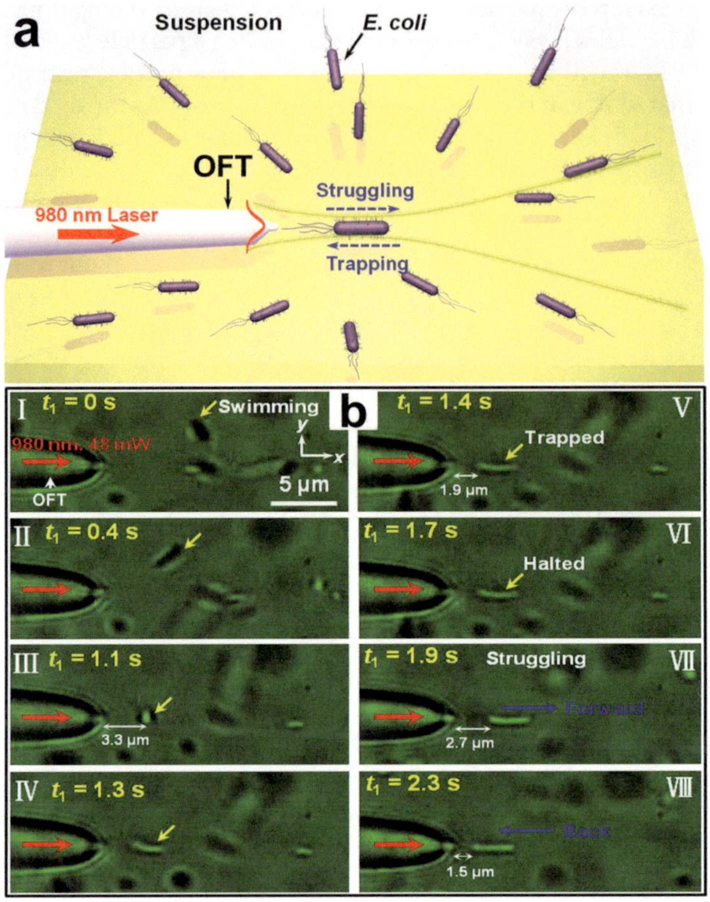

Figure 2.
Optical trapping of a single bacterium using OFTs [15]. (a) Schematic illustration of the non-contact optical trapping of a single bacterium and the struggling dynamics. (b) Optical microscope images of the trapping and struggling process of a single bacterium.

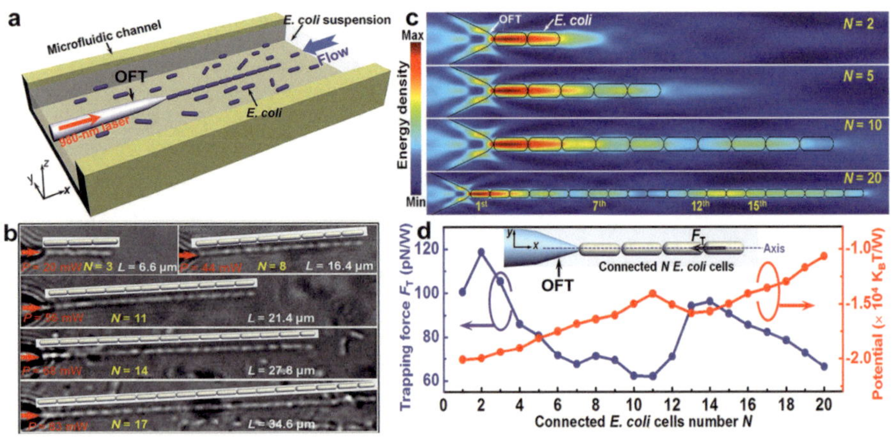

Figure 3.
Optical trapping of multiple cells using OFTs [16]. (a) Schematic of multiple E. coli trapping using OFTs. A laser at 980 nm wavelength was launched into the fiber probe which was placed in a microfluidic channel with a flowing suspension of E. coli cells. Multiple E. coli cells were trapped and connected orderly at the tip of the fiber probe. (b) Simulated light propagation along multiple bacteria. (c) Simulated light distribution along the assembled cell chains. (d) Calculated optical trapping force exerted on the last cell of each cell chain and the trapping potential.

4. Assembly of cell-based biophotonic waveguides by OFTs

Based on the multiple cell trapping capability of OFTs, direct formation of biophotonic waveguides with *E. coli* were reported [17]. By launching a laser of 980 nm wavelength into the OFTs, multiple *E. coli* were trapped and connected together with highly ordered organizations, forming biophotonic waveguides with different lengths (**Figure 4a**). By coupling a visible laser beam into the formed biophotonic waveguides, light propagation along these biophotonic waveguides can be directed observed as indicated by the red-light spots at the end of the waveguides (**Figure 4b**). The light propagation loss along the formed waveguides can be measured using an optical power meter by coupling another tapered optical fiber at the end of the formed biophotonic waveguide. As shown in **Figure 4c** and **d**, the measured propagation loss was measured to be 0.23 dB/μm.

In addition to the linear biophotonic waveguides, using OFTs, branched photonic probes can also been assembled. For example, **Figure 5** shows the assembled branched photonic probes with *E. coli* bacteria [18]. By designing a specially segmented tapered optical fiber, light output from the fiber can be divided into three individual beams, and *E. coli* bacteria can be trapped by the individual beams, further forming into branched biophotonic probes with different lengths (**Figure 5**). These branched photonic probes show strong stability, and can be used for further applications. By moving the OFTs, the formed biophotonic probes can be flexibly manipulated to different designated positions for further applications. These results show that the OFTs offer a seamless interface between optical and biological worlds for biophotonic probes formation with natural

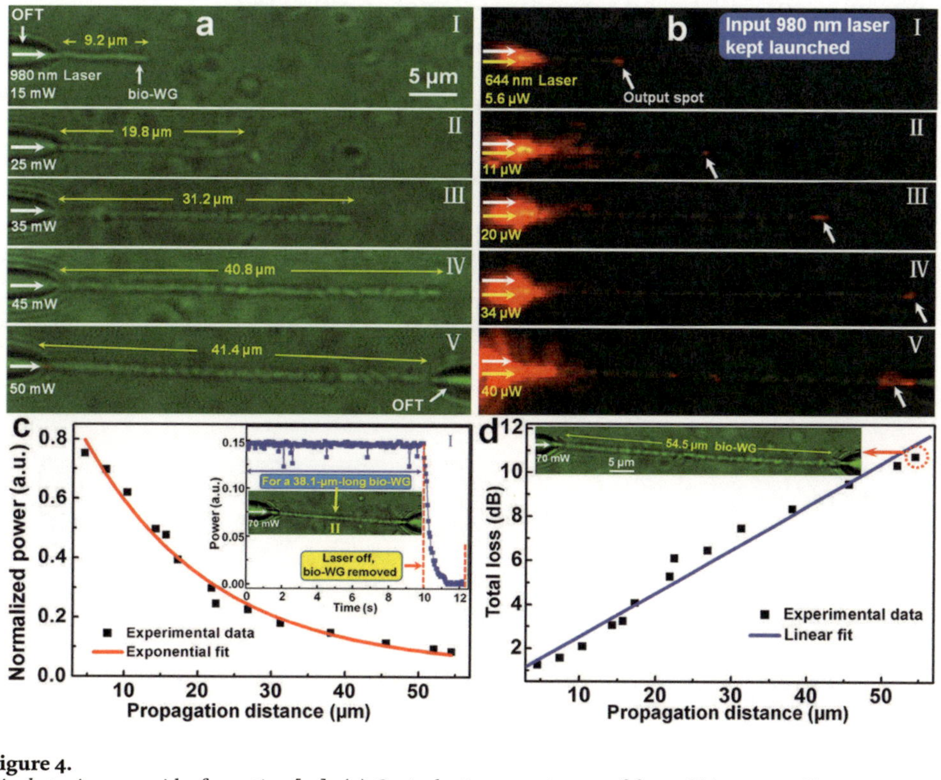

Figure 4.
Biophotonic waveguides formation [17]. (a) Optical microscope images of formed bio-waveguides (bio-WGs) with different lengths. (b) Light propagation observation along the formed biophotonic waveguides. (c) Normalized optical power measured at the end of each waveguides. (d) Measured optical loss of the waveguides.

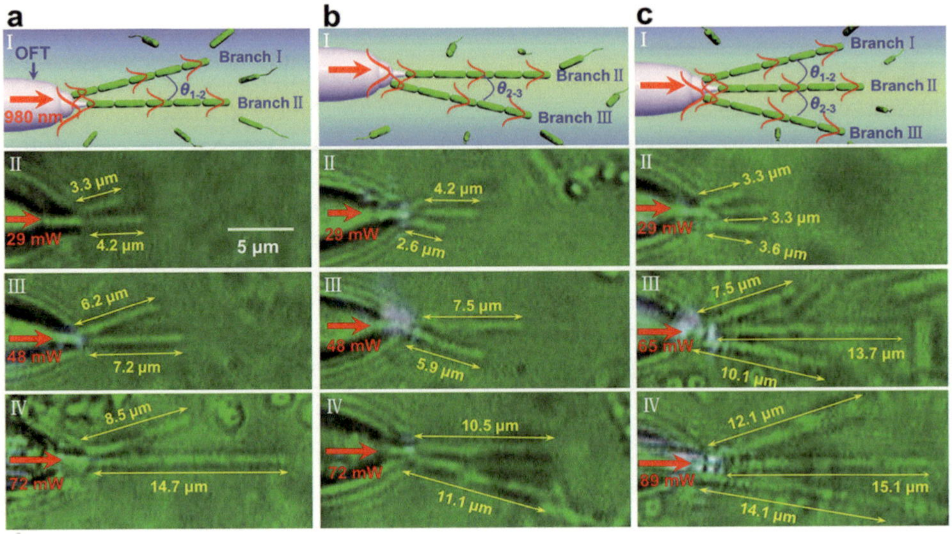

Figure 5.
Optical assembly of branched biophotonic structures [18]. (a, b) Assembly of two-branch structures. (c) Assembly of three-branch structures.

materials, and provides a new opportunity for direct sensing and detection of biological signal and information in biocompatible microenvironments.

5. Assembly of cell-based periodical structures by OFTs

In addition to the assembly of biophotonic waveguides with one type of cells, assembly of periodical structures of different types of cells was also demonstrated using OFTs [19]. Using *E. coli* cells and *Chlorella* cells as examples, different cells are flexibly patterned into one-dimensional (1D) periodic cell structures with controllable configurations and lengths (**Figure 6**), by periodically connecting one type of cells with another by optical force. Further demonstration shows that the structures show good performance for light propagation and can be moved flexibly. Real-time light signals can be detected from these photonic structures. These features make these photonic structures excellent candidates for the detection of signals transducing among different patterned cells. This assembly and patterning technique can also be applicable for other cells, such as mammalian cells and human cells.

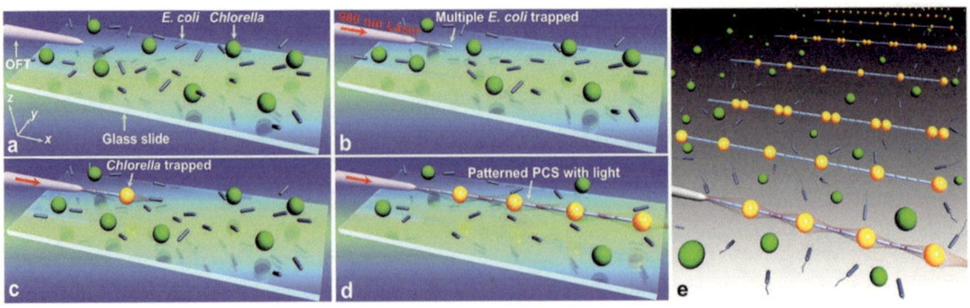

Figure 6.
Experimental schemes for cell assembly into periodical structures [19]. (a) an OTF is placed in cell suspensions. (b) Laser launched, multiple E. coli cells trapped. (c) a Chlorella cell is trapped and connected to the former trapped E. coli cells. (d) a periodical structure is formed, and light propagates along the periodical structure. (e) Schematic shows the assembled periodical biophotonic structures.

6. Assembly of cell-based structures in vivo by OFTs

The assembly capability can also be used for in vivo applications. For example, a non-contact intracellular binding and controllable manipulation of chloroplasts *in vivo* was demonstrated using OFTs [12]. By launching a laser beam at 980 nm wavelength into the tapered fiber, which was placed above the surface of a living plant (*Hydrilla verticillata*) leaf with a gap of about 3 μm to the leaf surface, chloroplasts with different numbers were stably bound and arranged into one-dimensional chains and two-dimensional arrays inside the leaf by optical force without damage to the chloroplasts, by the cooperation of scattering force F_s and gradient force F_g (**Figure 7**). The formed chloroplast chains were controllably transported inside the living cells. This non-invasive and non-contact method of organelle binding and manipulation could provide a way for biological and biochemical research *in vivo*, especially for investigating signal transduction and communication between intracellular organelles via organized organelle-organelle contact.

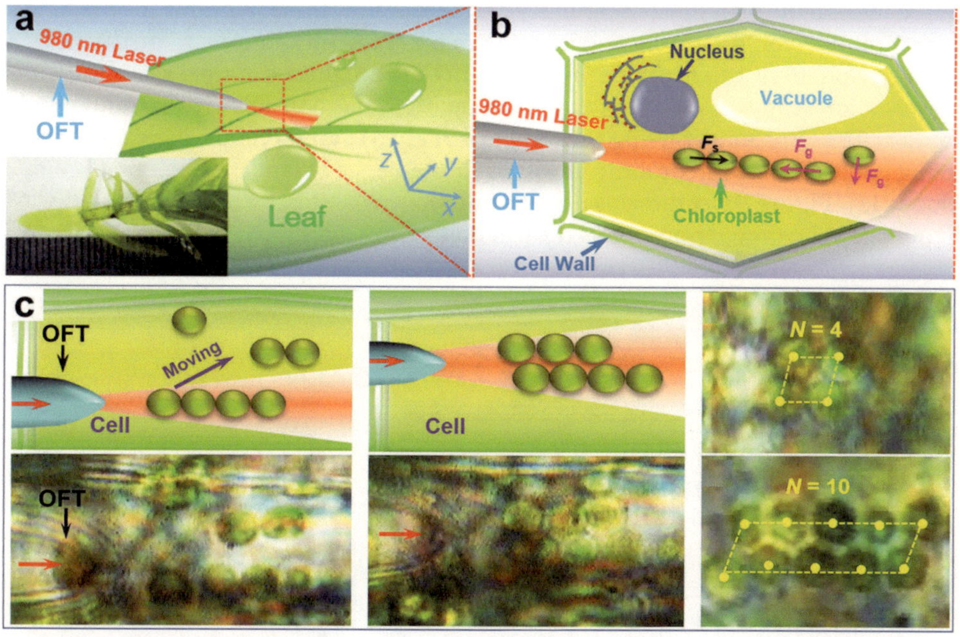

Figure 7.
Assembly of biophotonic probes in vivo [12]. (a) Schematic illustration of biophotonic probe assembly inside a leaf using OFTs. (b) Schematic illustration of biophotonic probe assembly based on a chain of chloroplasts. The chloroplasts inside a leaf are trapped and assembled by the cooperation of F_g and F_s. (c) Schematics and microscope images of the manipulation and assembly of organelle-based biophotonic probes in vivo.

7. Assembly of living photonic probe by OFTs for bio-probing and detection

Recently, using OFTs, a fully biocompatible living photonic probe for subwavelength probing of localized fluorescence from leukemia single-cells in human blood has been created [13]. The high-aspect-ratio living photonic probe based on a yeast cell (1.4 μm in radius) and Lactobacillus acidophilus (*L. acidophilus*) cells (2 μm in length and 200 nm in radius) is formed at the tip of a tapered optical fiber by optical trapping (**Figure 8a**). In the assembly, the authors have precisely moved the fiber to approach a yeast cell. Benefited from the spherical shape of the yeast, the trapping laser beam was focused into a tiny region and exerted a strong optical force on a *L. acidophilus* cell that traps it behind the yeast. With this alignment, the trapping laser beam propagates through the *L. acidophilus* cell and exert an optical force on other *L. acidophilus* cells, which were orderly bound together by optical binding effect and finally formed the living photonic probe. **Figure 8b** shows a formed probe assembled with a yeast and five *L. acidophilus* cells. To view the light propagation, after assembly of the probe, the trapping laser remained on, and a visible illumination light was launched into the probe. **Figure 8c–e** show the illumination light propagating along the tapered fiber. At the output port of the probe, a tiny light spot was observed with full width at half maximum (FWHM) of 345, 282, and 248 nm for the illumination wavelengths of 644, 532, and 473 nm, respectively.

As a benefit of the highly focused effect of the living cells, the living photonic probe can also deliver subwavelength excitation light to biological samples, and detect optical signals with a subwavelength spatial resolution. Moreover, within human blood, selective probing of the localized fluorescent signals on single leukemia cell surface can be realized via the precise manipulation of the living photonic probe. Due to the high biocompatibility and resolution, these photonic probes hold great promises for biosensing and imaging in bio-microenvironment. Furthermore, the living photonic probe can be integrated in the available near-field scanning optical microscopy, functioning as a biocompatible and non-invasive scanning probe for near-field imaging of living cells. **Figure 9**, as an example, shows the use of the living photonic probe in probing localized fluorescence of leukemia cells in human blood [13]. **Figure 9a–d** shows the spot excitation capability by manipulating the living photonic probe to approach the cell membrane. As shown in **Figure 9a**,

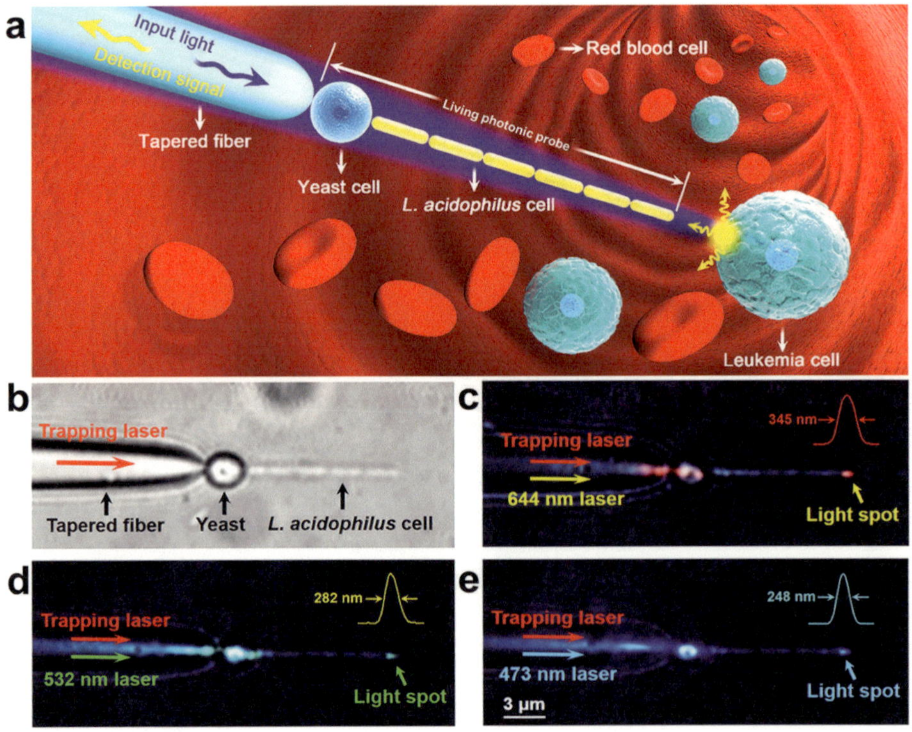

Figure 8.
Assembly of living biophonic probes for bio-probing [13]. (a) Schematic illustration for assembly of living photonic probe by OFTs. (b) Image of a formed living photonic probe. (c)-(e) images showing light propagation along the formed living photonic probes. Light spots can be observed at the end of each photonic probes.

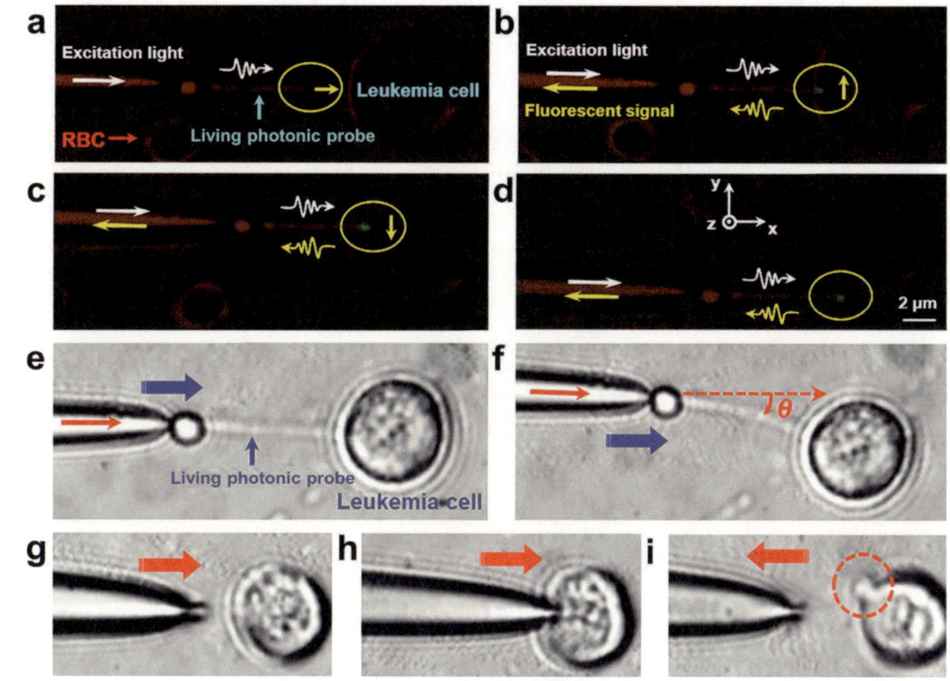

Figure 9.
Living photonic probe for single-cell probing and detection [13]. (a-d) Excitation and detection of local fluorescence from a leukemia cell in human blood by manipulating the living photonic probe to scan a cell. (e,f) Flexibility testing of the probe by pushing the probe against the leukemia cell membrane. (g-i) Touching and punching of the cell directly using a tapered optical fiber tip, to compare the flexibility of the living photonic probe.

there was no fluorescent signals when the distance between the living photonic probe and the surface of a leukemia cell was 3 μm. But the fluorescent signal was detected with a distinct fluorescent spot observed at the cell membrane when the probe was in contact with the cell (**Figure 9b**). The fluorescent signals at other locations were also detected by scanning the cell surface via precisely moving the probe (**Figure 9c** and **d**). Flexibility and deformability of the living photonic probe have also been demonstrated by interacting with biospecimens. As shown in **Figure 9e** and **f**, the living photonic probe was forced against a leukemia cell, then the living photonic probe was bent to an angle θ of 15° without puncture to the cell membrane. A certain degree of the deformability of the probe has no obvious influence on the scanning capabilities. For comparison, the authors pushed a fiber probe with a sub-micrometer tip, which is commonly used in scanning probe microscopes, against the leukemia cell (**Figure 9g**). As a result of the relatively large dimension and rigid structure, the fiber probe could easily insert into the cell (**Figure 9h**), and rupture the cell membrane (**Figure 9i**).

8. Conclusions

In this chapter, we reviewed the trapping and assembly of biological cells using OFTs, and finally extended the trapping capability for the assembly of living photonic probes such as cell-based biophotonic waveguides, cell-based periodical structures, cell-based structures in vivo, and living photonic probe for bio-probing and detection. These living photonic probes exhibit extremely high biocompatibility for further biological applications in bio-environment. As a benefit of the light focusing ability of the cells, the biocompatible living photonic probes allow the trapping, manipulation, sensing, and diagnostics in vivo. Furthermore, the living photonic probes assembled using OFTs offer an biophotonic bridge between optical and biological worlds with natural materials. With the advantages of its biocompatibility, the living photonic probes are envisioned to provides a new opportunity for direct sensing and detection of biological signal and information in biocompatible microenvironments.

Acknowledgements

This work was supported by the National Natural Science Foundation of China (No. 61975065, 11904132), Guangdong Basic and Applied Basic Research Foundation (2019B151502035), and Science and Technology Program of Guangzhou (202102010088).

Conflict of interest

The authors declare no competing financial interests.

Author details

Xing Li and Hongbao Xin*
Institute of Nanophotonics, Jinan Universtiy, Guangzhou, China

*Address all correspondence to: hongbaoxin@jnu.edu.cn

References

[1] Ashkin A, Dziedzic JM, Bjorkholm J. Observation of a single-beam gradient force optical trap for dielectric particles. Opt. Lett. 1986;11:288-290. DOI: 10.1364/OL.11.000288

[2] Ashkin A, Dziedzic JM. Optical trapping and manipulation of viruses and bacteria. Science. 1987;235:1517-1520. DOI: 10.1126/science.3547653

[3] Zhao X, Zhao N, Shi Y, Xin H. Optical fiber tweezers: a versatile tool for optical trapping and manipulation. Micromachines. 2020;11:114. DOI: 10.3390/mi11020114

[4] Fuh MRS, Burgess LW,Hirschfeld T. Single fiber optic fluorescence pH probe.Analyst. 1987;112:1159-1163. DOI: 10.1039/an9871201159

[5] Hu Z, Wang J, Liang J. Manipulation and arrangement of biological and dielectric particles by a lensed fiber probe. Opt. Express. 2004;12:4123-4128. DOI: 10.1364/OPEX.12.004123

[6] Lyons ER, Sonek GJ. Confinement and bistability in a tapered hemispherically lensed optical fiber trap. Appl. Phys. Lett. 1995;66:1584-1586. DOI: 10.1063/1.113859

[7] Taguchi K, Ueno H, Hiramatsu T. Optical rapping of dielectric particle and biological cell using optical fiber. Electron. Lett. 1997;33:413-414. DOI: 10.1049/el:19970247

[8] Taguchi K, Ueno H, Ikeda M. Rotational manipulation of a yeast cell using optical fibers. Electron. Lett. 1997;33:1249-1250. DOI: 10.1049/el:19970827

[9] Taguchi K, Atsuta K, Nakata T. Levitation of a microscopic object using plural optical fibers. Opt. Commun. 2000;176:43-47. DOI: 10.1016/S0030-4018(00)00499-5

[10] Frick M, Haller T, Dietl P. Combined optical tweezers and optical stretcher in microscopy. SPIE. 2001;4434:227-232. DOI: 10.1117/12.446684

[11] Ribeiro RSR, Soppera O, Oliva AG. New trends on optical fiber tweezers. J. Lightwave Technol. 2015;33:3394-3405.

[12] Li Y, Xin H, Liu X. Non-contact intracellular binding of chloroplasts in vivo. Sci. Rep. 2015;5:10925. DOI: 10.1038/srep10925

[13] Li Y, Xin H, Zhang Y. Living nanospear for near-field optical probing. ACS Nano. 2018;12:10703. DOI: 10.1021/acsnano.8b05235

[14] Xin H, Xu R, Li B. Optical trapping, driving, and arrangement of particles using a tapered fiber probe, Sci. Rep. 2012;2:818. DOI: 10.1038/srep00818

[15] Xin H, Liu Q, Li B. Non-contact fiber-optical trapping of motile bacteria: dynamics observation and energy estimation. Sci. Rep. 2014;4:6576. DOI: 10.1038/srep06576

[16] Xin H, Zhang Y, Lei H. Optofluidic realization and retaining of cell-cell contact using an abrupt tapered optical fiber. Sci. Rep. 2013;3:1993. DOI: 10.1038/srep01993

[17] Xin H, Li Y, Liu X. Escherichia coli-based biophotonic waveguides. Nano Lett. 2013;13:3408. DOI: 10.1021/nl401870d

[18] Xin H, Li Y, Li B. Bacteria-based branched structures for bionanophotonics. Laser Photonics Rev. 2015;9:554. DOI: 10.1002/lpor.201500097

[19] Xin H, Li Y, Li B. Controllable patterning of different cells via optical assembly of 1D periodic cell structures. Adv. Funct. Mater. 2015;25:2816. DOI: 10.1002/adfm.201500287

Chapter 7

Application of Fiber Optics in Bio-Sensing

Lokendra Singh, Niteshkumar Agarwal, Himnashu Barthwal, Bhupal Arya and Taresh Singh

Abstract

The unique properties of optical fibers such as small size, immunity to electromagnetic radiation, high sensitivity with simpler sensing systems have found their applications from structural monitoring to biomedical sensing. The inclusion of optical transducers, integrated electronics and new immobilization methods, the optical fibers have been used in industrial process, environmental monitoring, food processing and clinical applications. Further, the optical fiber sensing research has also been extended to the area of detection of micro-organisms such as bacteria, viruses, fungi and protozoa. The validation of optical fibers in bio-sensing applications can be observed from the growing number of publications. This chapter provides a brief picture of optical fiber biosensors, their geometries including the necessary procedure for their development. This chapter could be a milestone for the young researchers to establish their laboratory.

Keywords: optical fiber, biosensors, biomedical sensing, environmental monitoring, micro-organisms detection

1. Introduction

The inclusion of optical fibers in bio-sensing applications was started by two different, but interrelated discoveries, such as the laser light and optical fibers. The theoretical work of C. H. Townes and A. L. Schawlow was used by T. H. Maiman to develop the first laser. A optical signal obtained through laser is highly collimated, inherently coherent, and quasi monochromatic with the data transfer capability. The optical signal propagates in optical fiber by obeying the principle of total internal reflection (TIR) with very low losses and the first working model of optical fiber was proposed in 1965 [1]. The working model of optical fiber was put forwarded 100 years after the demonstration of concept of light. Since, then the main focus was to improve the transmission of optical signal through fibers. Nowadays, the key focus is on long distance high speed communication with low transmission losses such as 2 dB/km [2]. The unique properties of optical fibers such as immunity to electromagnetic (EM) interference and miniature footprints, the optical fiber has found niche application in sensing [3].

A schematic of conventional single mode fiber (SMF) used in the field of telecommunication is shown in **Figure 1**, consisting of three layers such as a silica core having diameter of in order of several microns ($\sim$ 2–9 μm) and doped with germanium to boost up its refractive index (RI), a silica cladding of diameter of 125 μm

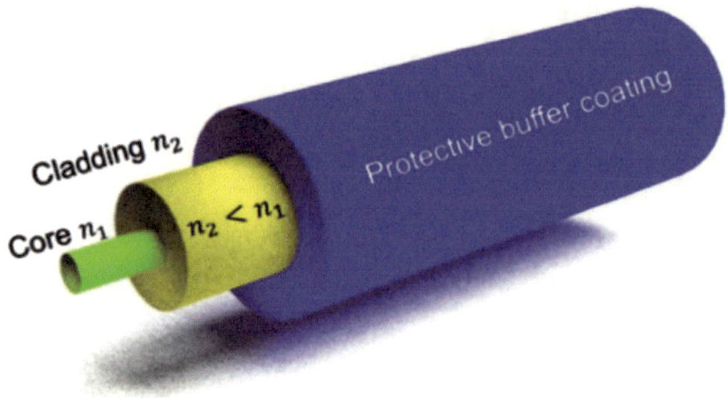

Figure 1.
Schematic of single mode optical fiber [1].

and a coating of plastic jacket. Although, the plastic coating does not play any role in light propagation but provides the mechanical strength to the fiber. The optical fibers can be fabricated by using some other materials such as chalcogenide [4], plastic [5], and composites, with different composite materials in core and cladding. Based on the core size, operating wavelength, and RI difference of core and cladding, an optical fiber can work in the regime of single or multimode. In single mode fibers, the distribution of optical signal profile in core is Gaussian, while in multimode signal profile is more complex [2].

The optical sensors detect the variation in optical properties of propagating signal, that occurs due to the physiochemical change in targeted environment. The optical fiber based sensors classified into two categories on the basis of sensing region such as extrinsic or intrinsic sensors. The sensors directing or collecting optical signal to and from external environment are termed as extrinsic sensors [6]. The sensors in which the properties of optical signal vary within the fiber are known as intrinsic sensors [7]. In general, extrinsic sensors being used for the detection of external stimuli such as physical or biochemical parameters. The optical fiber based measurement techniques have received a great attention especially in the field of structural monitoring, railway and aerospace, chemical and biological sensing, medical diagnosis and environmental monitoring.

Since, the key application of SMF were in the field of telecommunication, and hence, fabricated in such a way that the influence of external field can be minimized on propagating signal. However, for the efficient operation of optical fiber sensors, the interaction of optical signal with external environment should need to be maximized. This can be attained by adopting different optical fiber processing schemes which frequently utilizes the interaction of leaking fields with external environment. The commonly used geometry of optical fiber in sensing applications are discussed in following subsections.

1.1 Cladding less evanescent based optical fiber sensors

The easiest way to increase the interaction of evanescent waves (EW) with external medium is removal of cladding, and a schematic of cladding less optical fiber sensor is illustrated in **Figure 2**. The changes in propagation of optical signals due to variation in external environment facilitates the EW spectroscopy [8]. The facilitation of EW spectroscopy is highly sensitive and powerful technique to quantitatively and qualitatively investigate the environment present in the vicinity of sensing region of sensor. The EW leaks from core to cladding and the distance

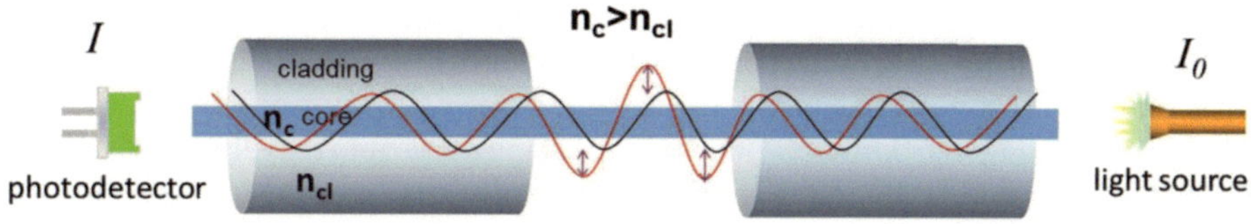

Figure 2.
Schematic of cladding less optical fiber sensor structure [3].

is termed as penetration depth. The penetration depth of EW can be evaluated as [9]:

$$d_p = \frac{\lambda}{2\pi\left(n_{eff}^2 - n_s^2\right)^{1/2}} \tag{1}$$

where, λ is the propagating wavelength, n_s is the RI of surrounding environment and n_{eff} is the RI of guided mode propagating in the core.

The absorption spectrum of surrounding medium attenuates the EW which hindered the propagating mode. This can be understood from Lambert–Beer Law which is given as:

$$\frac{I}{I_0} = c^* \alpha^* L \tag{2}$$

where, c is the concentration of absorption substance, $\boldsymbol{\alpha}$ is the attenuation constant of EW, and L is the path length in which optical signal interacts with the surrounding medium. I_0 and I are the intensities of the optical signal before or after the interaction to the external environment, respectively. The optical fiber sensor structure presented in **Figure 2** can be attained by removing the cladding part by using conventional approach such as treating the fiber with hydrofluoric (HF) acid [10]. To remove the cladding, fiber structure should need to be immersed in HF acid at constant stirring at 50 rpm. In cladding less optical fiber sensors the interaction of optical signal with surrounding can be enhanced by bending it in U-shape [11]. The U-shape bend is also useful for monitoring because source and detector will be on same side. Although, the cladding less fiber can also be attained by using other techniques such as plasma etching, but it will turn into expensive systems.

1.2 Tapered optical fiber sensor

An access to EW can also be obtained by tapering the optical fiber structure. The tapering of optical fiber usually done within the dimensions varying from

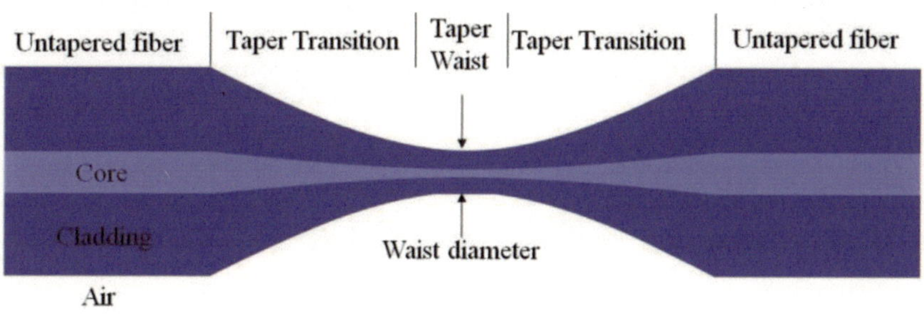

Figure 3.
Schematic of tapered optical fiber structure [12].

submillimeter to several millimeters. The tapered region of the optical fiber maintains the uniform diameter with conical ends to merge it with unaltered part of optical fiber as illustrated in **Figure 3**. The tapering of fiber is done by heating the fiber structure by using flame or CO_2 laser beam. The properties of tapered optical fiber sensor is based on the diameter of conical ends, diameter of tapered region, and RI of surroundings. The proportion of EW power in tapered fiber structure, increases with decrease in diameter of tapered region and with decreasing RI difference of external environment and of fiber [13]. The tapered optical fiber provides numerous advantages to the sensors such as compactness, higher sensitivity and flexibility. The tapered optical fiber classified into categories such as adiabatic and non-adiabatic. When the tapered transition region is small in such a way that maximum optical power confines within the core, then such structure are termed as adiabatic tapered fibers [13]. However, in non-adiabatic one the diameter of tapered region is less than 10 μm and the propagating modes couples into higher order modes [14]. The tapered optical fibers have been utilized in various sensing applications [15–17]. In case of tapered fiber structures, the interaction of EW with surrounding medium can be analyzed by two different.

approaches. In first approach, the attenuation of signal is to be measured which is propagating through tapered region and depends on the RI of surrounding medium [18]. In second one, the variation in surrounding medium affects the RI of modes propagating in the tapered section of fiber and works interferometrically, by using mode theory [19].

1.3 Interferometers

The optical fiber interferometers provide very high sensitivity because of their unique operational mechanism and usually known as modal interferometers (MI). In MI basically, the propagating modes splits into two modes at sensing region which are traveled in different RI regime that causes a difference in their phase and wavelength. The different properties of propagating modes lead to the interference in fundamental and higher order modes and results into a transmission spectrum with fringes. The phase of the fringes ca be given as:

$$\varphi = \frac{2\pi}{\lambda}(\delta n_{eff})L \tag{3}$$

where, L is the center to center distance between two modes and λ is the operating wavelength [20]. A SMF and thin core fiber (TCF) based Mach-Zehnder interferometer (MZI) is presented in **Figure 4**. The first strand of SMF carries a single mode which splits into two parts at TCF due to variation in core diameter. In second strand of SMF the modes from TCF gets recombined at SMF. The difference in phase of recombined modes leads to the addition or cancelation of phase at output of MZI [20].

The optical fiber based Michelson interferometers were also proposed and a schematic is illustrated in **Figure 5**. In Michelson interferometer, the core modes

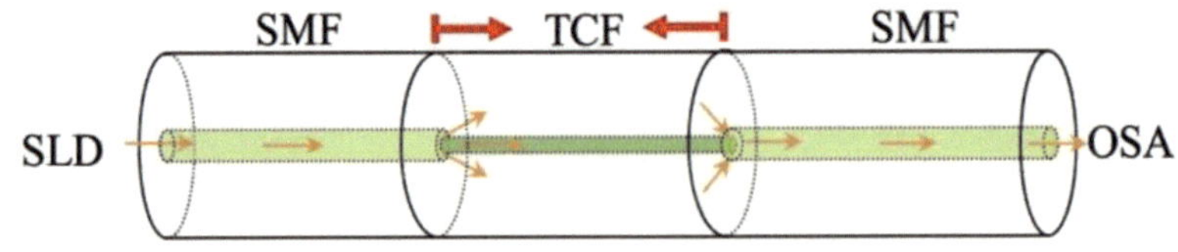

Figure 4.
Schematic of SMF and TCF fiber based Mach-Zehnder interferometer [21].

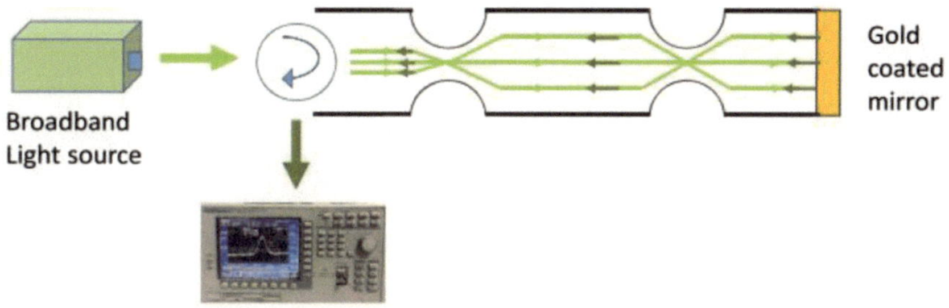

Figure 5.
Schematic of optical fiber based Michelson interferometer [3].

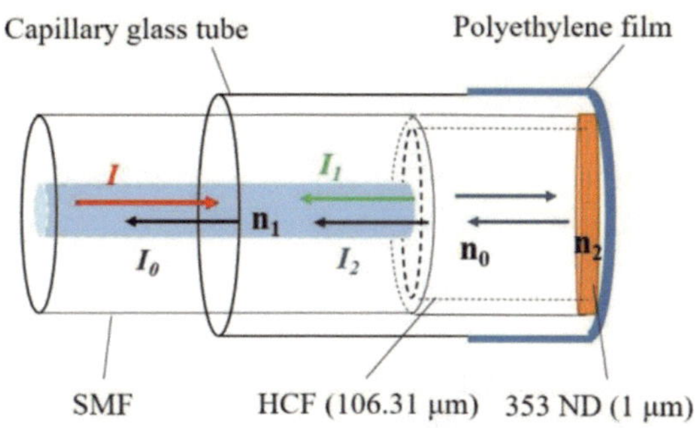

Figure 6.
Schematic of optical fiber based FPI sensor [22].

distributed into higher order modes at tapered section and after striking to gold film reflects back and recombined at the tapered section. Therefore, an interference between the modes occurs at the tapered region that causes the generation of fringes. The presence of external medium in the region separating the taper and gold films introduces the interfering features in the received signal. In similar physical length, Michelson interferometer provides higher sensitivity because the twice interaction of optical signal with sensing region. These interferometers works on the basis of measurement of wavelength or amplitude of the spectrum.

received at the output. An another type of optical fiber based interferometer is Fabry-Perot interferometer (FPI). The FPI is consisting of a cavity between two reflectors and illustrated in **Figure 6**. Alternatively, a FPI can be developed by coating a thin metallic layer at the tip of the fiber which acts as a mirror and the distance between metallic layer and surrounding medium as an another mirror. A change in the RI of cavity or its length can modulate the signal. The modulated signal will be further used to measure the targeted measurand that modulates the signal.

1.4 Grating based optical fiber sensor

An optical fiber grating is consisting of slots placed periodically with an equal proportion. The slots in optical fiber structure leads to the modulation of the propagating optical signal. The grating can be incorporated by exposing the fiber structure to the ultra-violent or femtosecond laser with desired geometry [23]. The optical fiber based grating structure were also found to be a good candidate for the sensing applications [23]. A schematic of FBG sensor with its measurement setup is illustrated in **Figure 7**. The grating structure couples the forward and backward propagating modes of the core at the particular wavelength that satisfies the Bragg condition. A Bragg grating is considered as reflector which reflects a specific wavelength band

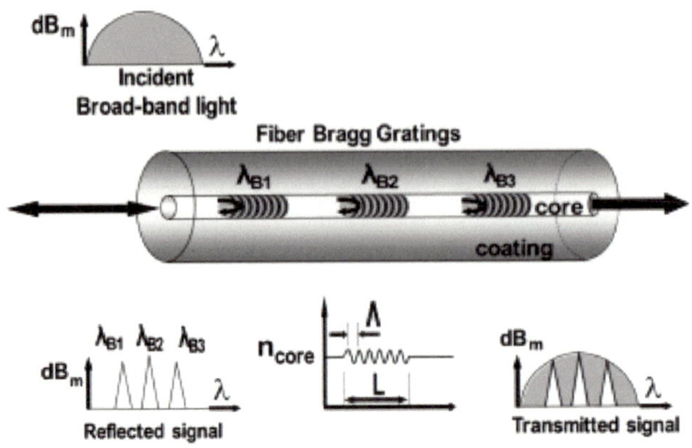

Figure 7.
Schematic of measurement setup of FBG sensor [24].

along the optical fiber and transmitted all others. The reflected Bragg wavelength is governed by a mathematical expression which can be given as [23]:

$$\lambda_{\text{Bragg}} = 2\eta_{\text{eff}} \quad (4)$$

In Bragg grating based sensors, the interaction of EW with surroundings can be maintained or enhanced by modifying the fiber geometry such as tapering, etching of cladding of sensing region. Therefore, to overcome this limitation, tilted Bragg grating can be utilized in which the gratings are designed at a specific angle with respect to the axis of the core. The interaction of cladding modes with EW changes the wavelength of propagating cladding modes [25]. The interaction of EW with surrounding medium leads to the induction of inherent sensitivity to the external RI and to the nano-coatings placed over the cladding layers. While considering the fact long periodic (LPG) grating structures were come into origin. The LPG are generally created with in the length of 100 microns to 1 mm as illustrated in **Figure 8**. LPG usually couples the light form the core modes to the co-propagating modes of the structure [27]. The cladding mode suffers higher attenuation, therefore, the transmission spectrum of LPG can be analyzed by using the series of resonance bands.

From the above discussion, it can be concluded that optical fiber based sensors have wide applications in bio-sensing applications. A short summary of above discussed different geometries of optical fiber sensor structures is tabulated in **Table 1**. The tabulated form is easy enough to get a brief introduction to the required geometry of sensor.

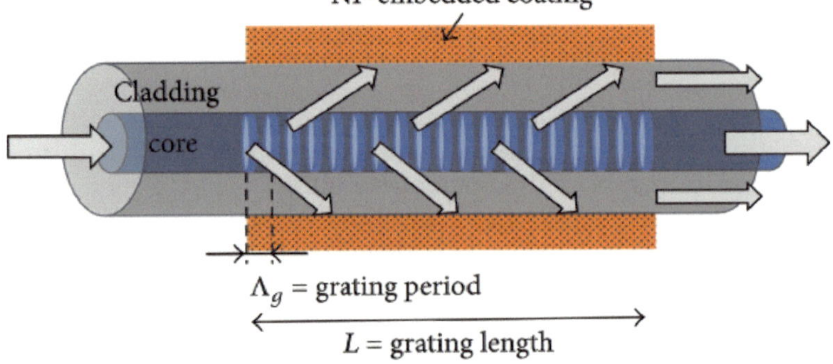

Figure 8.
Schematic of NP coated LPG sensor structure [26].

Sensor type	Measurand	Light parameters	Units
Cladding less	Absorption, concentration	Intensity	dB, %
Tapered	RI, absorption, concentration, pressure, temperature, strain	Wavelength shift, intensity	dB, %, nm
Interferometers	RI, absorption, concentration, pressure, temperature, strain	Intensity, wavelength shift, phase	dB, %, nm, degrees
FBG	pressure, temperature, strain	Intensity, wavelength shift	dB, %, pm
LPG	RI, absorption, concentration, pressure, temperature, strain	Intensity, wavelength shift	dB, %, nm

Table 1.
Summary of measurand and light parameters of different sensor structures.

2. Biochemical measurands in healthcare

The optimum properties of optical fibers such as higher sensitivity and low limit of detection are the crucial parameters, but in addition, the selectivity is also an important concept in biochemical measurement. The selectivity or specificity is important to avoid the interference of other biomolecules or biomarkers presented in targeted analytes. There are two approaches based on which the selectivity of biosensor can be attained. The first approach is to use special material fibers such as chalcogenide glasses, fluoride or silver halide glasses [28]. These fibers are transparent to IR wavelength, and on the contrary, biomolecules pursue the highly absorption features [29, 30]. However, the use of chalcogenide fibers is not useful because of their potential toxicity and still an effort is required to improve their responses towards biomolecules [28]. In second approach, there is indirect sensing of analytes by placing a biochemical layer over the sensing region. The biochemical layer changes the optical properties on the basis of surrounding RI. Such biosensors provide the quantitative and qualitative information of the chemical reagent under examinations. The chemical layer over the sensing region means the wavelength of output optical signal is managed by the properties of biochemical layer instead of absorption spectra. The sensitivity of such biosensors is depends on the length of sensing area, amount of EW and optical properties of the coated biochemical layer [31].

2.1 Chemical optical Fiber sensors

The diagnosis of biomolecules present in human bodies can be detected in two phase such as in gases or in liquid. In gas phase, the analysis can be done by analyzing the gases exhaled from skin or breath. In liquid phase, the analysis of biomolecules can be done by testing the samples such as urine, saliva, blood, sweat and tears.

2.1.1 Diagnosis in gas phase

The biomarkers released from human bodies are useful to develop the non-invasive techniques. The diagnosis of these biomarkers is important to find the presence of disease [32, 33]. The breath sniffing method is useful to analyze the patient suffering from renal failure in rats [34] and lung cancer detection [35]. Oxygen and carbon dioxide are the two gases that are routinely checked in clinical applications. The detection of these two gases was also performed by using optical fiber sensor by using pH indicator separated with well separated with emission bands [36]. Ammonia is one of the major component that affects the body metabolism and can disturb the functioning of kidney and liver [37, 38]. In normal conditions, the ammonia releases from body skin from slight alkaline blood and its detection is used to diagnose the disease related to kidney and liver [39]. The ammonia diagnosis was carried out by using optical fiber sensors. Initially, the detection was done by employing pH detector based on indications [40]. Since then, reflector sensor tips [41], EW based fiber grating [42], and lossy mode resonance (LMR) [43] were reported. The sensitivity and limit of detection of such optical fiber sensor was extremely good in comparison of existing works.

The diagnosis of various organic compound is hardly done at clinical level, but number of studies were reported. Although, the optical fiber sensors for the detection of organic compounds are not very sensitive [44]. An EW based optical fiber sensor was put forwarded for the detection of gas exhaled from human skin [45]. The proposed sensor is also capable of analyzing the physiological changes by applying a pattern recognition technique. The optical fiber sensors have also been utilized for the diagnosis of humidity, which is one of the important factor in case of critical conditions [46]. The increase in humidity in human bodies leads to the dryness in mucosa and cause difficulties in breathing. However, instead of such critical need, the optical fiber based humidity sensors cannot be used in medical applications because of slow response and recovery time.

2.1.2 Diagnosis in liquid phase

The diagnosis of biomarkers present in human bodies can be done by measuring the pH of liquid. The pH of liquid present in stomach is varies from 1.3 to 3.5, and of urine and pancreas is from 8.0 to 8.8 [47]. A tilted FBG based sensor structure was reported to detect the pH of human body fluids [48]. The sensor is working on the basis of coated polymer films whose thickness varies according to the variation of body fluid concentration and leads to the change in optical properties of the signal. Despite of reported articles, the pH sensors have been utilized *in vivo* applications and are commercially provided by the Ocean Optics [49] and PreSens [50] with enough capabilities to be utilized in medical applications.

The pH detection in bio-fluids is also useful to detect the presence of drugs which will be helpful for pharmaceuticals and could be a milestone to develop therapeutic aids for human and animals [51]. The detection of antibiotics in human blood stream can be a useful step to prevent the overdose or to provide the effective dose for specific disorder. A LPG based vancomycin sensor was reported which can be used to treat some severe gram-positive infections [52]. The sensor is capable of detecting the very low concentration of antibiotics present in blood stream which were at the concentration of 10 nM with high specificity towards other biomolecules. Similarly, propofol is an anesthetic usually used in surgery and in regular use in intensive care units. Therefore, the detection of presence of propofol in human body is also an important factor, and a work was put forwarded for its detection while employing the optical fibers [53]. The reported work demonstrated a strong linearity with whole blood samples of human bodies.

3. Characterization and analysis process of optical fiber biosensors

The different geometries of optical fiber sensors should need to be characterized before involving them in sensing of biomolecules. The development of optical fiber biosensors involves four different process such as fiber geometries, used nanoparticles, detection of biomolecules and sensing analysis of developed sensor probes. Therefore, this section presents a brief discussion of about the necessary characterization of optical fiber sensors at all the steps.

3.1 Optical fiber sensor geometry

The validation of drawn fiber sensor geometry such as tapered fiber, MZI etc. can be done by using scanning electron microscopy (SEM). SEM is a kind of electron microscopy which employs a focused beam of the electron to analyze the surface of optical fibers. The SEM image of a tapered optical fiber probe is illustrated in **Figure 9**. In **Figure 9**, there are two SEM images where the first one is representing an image of tapered optical fiber sensor and another image such as **Figure 9(b)** is representing the distribution of nanoparticles coated over the sensing region of fiber structure. In some other cases, the diameter analysis of tapered optical fiber sensor structure was measured directly by using the fabricating machine, but the accuracy of the measured diameter was not up to the mark, and illustrated in **Figure 10** [55].

3.2 Nanoparticles

The optical fiber sensor structures also utilize the immobilization of nanoparticles over the sensing region to enhance the sensitivity by means of introducing the concept of localized surface plasmon resonance (LSPR) phenomenon.

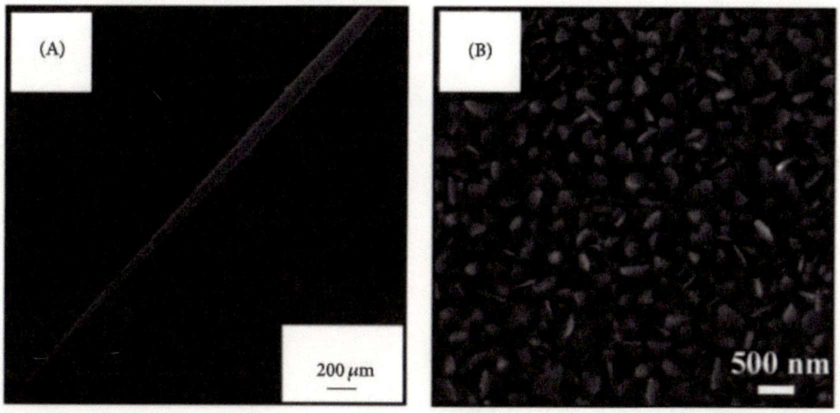

Figure 9.
SEM image of SERS probe of a tapered optical fiber sensor structure: (A) tapered optical fiber, and (B) distribution of nanoparticles over the fiber [54].

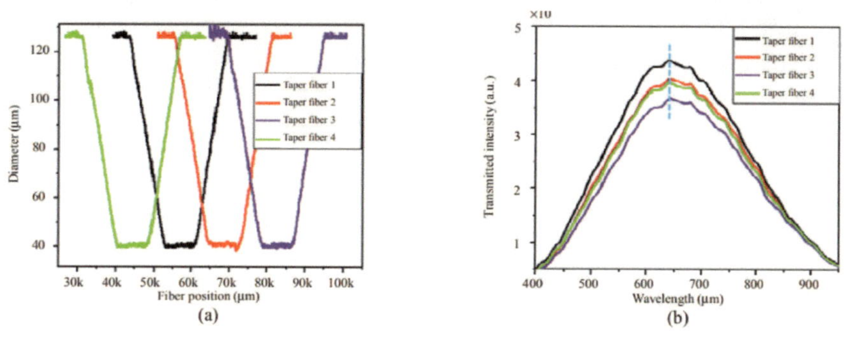

Figure 10.
Analysis of tapered optical fiber sensor structure: (a) diameter analysis, and (b) transmitted spectra [55].

The characterization of nanoparticles can be done by using UV-spectrophotometer and by observing their distribution through transmission electron microscopy (TEM) images. TEM is also a technique that employs a focused beam of electrons to visualize the distribution of particles in nanometer dimensions. The UV-spectrophotometer provides the resonance peak of the nanoparticles through the absorbance spectrum and is useful to confirm their initial synthesis. The resonance peak of all the nanoparticle is different and usually falls in the visible spectrum of white light. The peak resonance wavelength in absorbance spectrum of gold and zinc oxide nanoparticles appears at 519 nm and 370 nm for the particles size of less than 15 nm and 50 nm, respectively, and illustrated in **Figure 11**. The initial confirmation of nanoparticles can be carried forward to analyze their distribution which usually done by using capturing the microscopic image by using TEM. The TEM images of gold and zinc nanoparticles are illustrated in **Figure 12**. From the TEM images it can be concluded that the distribution of nanoparticles is uniform and easily visible. Further, the morphology of the nanoparticles or layered nanomaterials is also an important factor to assure the synthesis of nanoparticles,

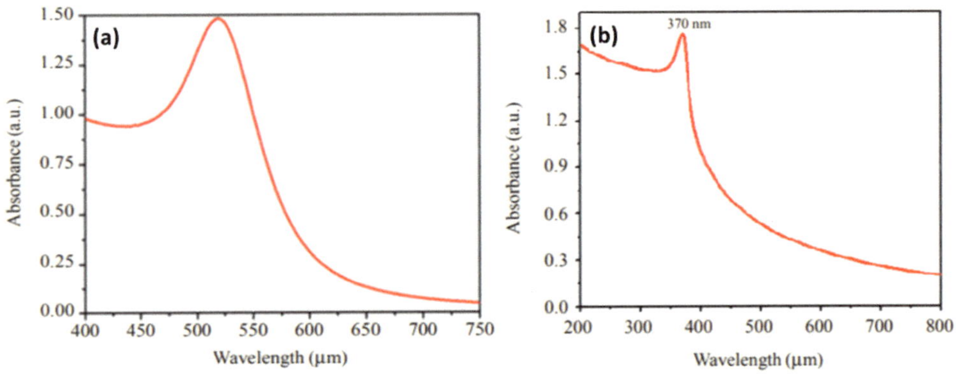

Figure 11.
Absorbance spectrum of nanoparticles: (a) gold, and (b) zinc oxide [55].

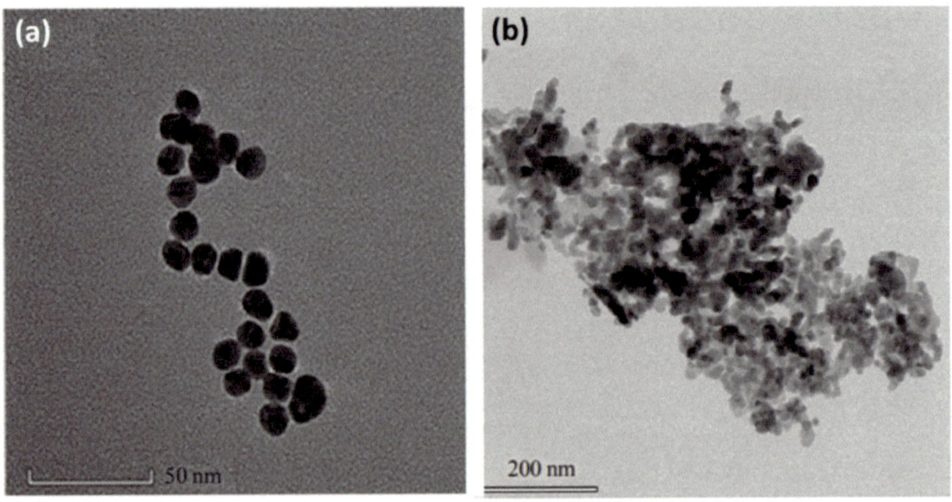

Figure 12.
TEM images of nanoparticles: (a) gold, and (b) zinc oxide [55].

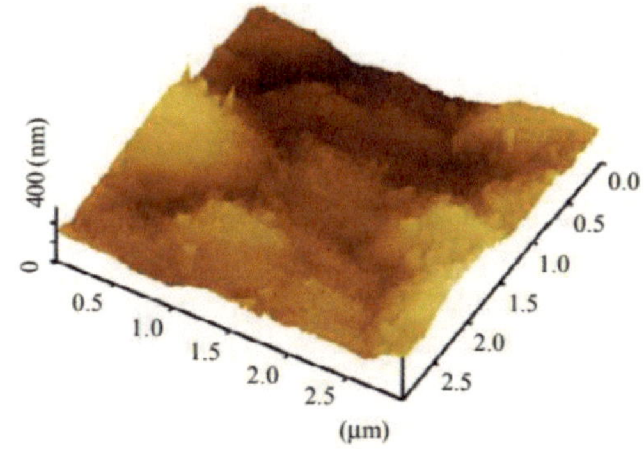

Figure 13.
AFM image of zinc oxide nanoparticles [55].

and can be done by taking the images by using atomic force microscopy (AFM). An AFM image of zinc oxide nanoparticles is illustrated in **Figure 13**.

3.3 Biomolecules

The preparation of samples of targeted biomolecules is also an important factor which helps in increase the performance of sensor probe. The analysis of samples of the targeted biomolecules can be done by preparing them in different pH base solutions. The similar kind of approach has been used to analyze the validity of ascorbic acid (AA) samples and illustrated in **Figure 14**. The performed test was basically done to check the solubility of artificial samples of AA [55]. The analysis was done by dissolving the artificial sample of AA in different pH solutions and the samples of lowest and highest concentration were prepared. Then, the peak resonance wavelength was measured for the highest and lowest sample concentration and their difference is plotted with respect to each pH solution. For the reported work, it was concluded that the AA samples are highly soluble in phosphate buffer solution (PBS) whose pH is about 7.4.

3.4 Sensing analysis

The sensing analysis of the sensor probe can be done in several steps. The first step is to sense all the samples through the sensor probe. For each measurement

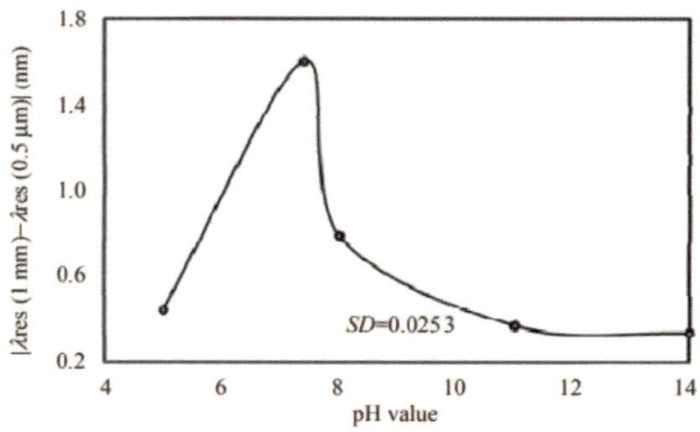

Figure 14.
Solubility test of ascorbic acid samples in different pH solutions [55].

respective peak resonance wavelength can be recorded which is useful to plot the autocorrelation coefficient of the sensor probe. The autocorrelation curve is used to evaluate the linearity, regression coefficient, sensitivity and resolution of the sensor. Then, the analysis of sensor can be done in terms of stability, reusability, reproducibility and selectivity.

The stability of any optical fiber biosensor can be evaluated by measuring the base solution through a sensor probe more than 10 times. The results can be plotted in terms of number of measurements and peak resonance wavelength. Then, the standard deviation (SD) can be evaluated to observe the stability and for a good sensor SD is usually less than 0.1.

The reusability is an another important parameter to analyze the performance of optical fiber sensor. Reusability can be evaluated by measuring two different concentration of bio-molecules through the same sensor probe. The measurement of any concentration should need to be performed three times to attain higher accuracy. The sensor head must need to be rinsed properly after all the measurements by using base solution. Then, the results can be plotted in terms of recorded spectra or in terms of peak absorbance wavelength. The resonance wavelength for similar concentration should be same for each measurement to attain the higher reusability.

The reproducibility is also an another important factor to analyze the performance of any optical fiber sensor. The reproducibility test can be done by measuring the similar concentration of bio-samples through one sensor probe. The measurement must need to be done at least 5 times to attain the higher accuracy. The outcome of the measurements can be plotted in terms of recorded spectra and in terms of peak resonance wavelengths. The higher reproducibility of the probe can be claimed if the peak resonance wavelength for all the measurements is similar.

The selectivity or specificity of the optical fiber sensor is a crucial factor of an optical fiber biosensor which helps in to remove the interference of other biomolecules present in real liquid samples of human bodies. The higher specificity of any optical fiber sensor can be attained by functionalizing the sensor head with appropriate enzyme which oxidize only in the presence of targeted bio-samples. For instance, the AA oxidized only in the presence of ascorbate oxidase.

4. Conclusions

This book chapter presents a brief discussion about the different optical fiber geometries which have been utilized for the development of different optical fiber sensors and biosensors. The mostly common used geometry of optical fibers is cladding less, tapered, interferometers, and gratings. The second section of the chapter presents the brief discussion about the presence of biochemical markers usually used in bio-sensing applications. The detection of biochemical markers is generally done in two phases such as in gas phase and in liquid phase. The third section of the chapter presents a brief discussion of the characterization and sensing process of the optical fiber based biosensors. The characterization of optical fiber sensor is done by capturing the images through TEM, SEM and AFM. The analysis of nanoparticles can be done by recording the absorbance spectrum by using UV-spectrophotometer. The sensing analysis of the optical fiber sensor can be done by performing the stability, reusability, reproducibility and selectivity test of the sensor probe. The optical fiber based biosensors are emerging in current era and can be employed in various health care applications.

Acknowledgements

On the behalf of all the authors, I Dr. Lokendra Singh would like to thank the Dr. S.P. Gupta, Hon'ble Vice Chancellor, University of Engineering and Technology Roorkee (UETR) and Dr. Brij Mohan Singh, Director, College of Engineering Roorkee for their continuous support and motivation while writing this book chapter.

Author details

Lokendra Singh[1*], Niteshkumar Agarwal[2], Himnashu Barthwal[1], Bhupal Arya[1] and Taresh Singh[1]

1 Department of Computer Science and Engineering, College of Engineering Roorkee (COER), Roorkee, Haridwar, Uttarakhand, India

2 Institute of Advance Research (IAR), Gandhinagar, Gujarat, India

*Address all correspondence to: kashyap00000@gmail.com

References

[1] Hecht J, City of Light, The Story of Fiber Optics (New York: Oxford University Press) 1999, p 114.

[2] Parries M C, Optical fiber Contemp. Phys. 1989; 30: 303–304. DOI: 10.1080/00107518908225520.

[3] Correia R, James S, Lee S-W, Morgan S P and Korposh4 S. Biomedical application of optical fiber sensors. J. Opt., 2018; 20: 073003.

[4] https://thorlabs.com/newgrouppage9.cfm?objectgroup. (Accessed July 29, 2021)

[5] https://thorlabs.com/catalogPages/1100.pdf. (Accessed July 29, 2021).

[6] Grattan K and Meggitt B. Chemical and Environmental Sensing. 1999; Boston, MA: Kluwer Academic.

[7] Mehvar M, BIS C, Scharer C M, Young M M and Luong J H. Fiber-optic biosensors-trends and advances. Anal. Sci., 2000: 16, 677–672.

[8] Korposh S, Kodaira S, Lee S-W, Batty W J and James S W. Nano-assembled thin film gas sensor: II. An intrinsic high sensitive fibre optic sensor for ammonia detection. Sensor Mater., 2009: 21, 179–189.

[9] Grattan K and Meggitt B. Chemical and Environmental Sensing, Boston, MA: Kluwer Academic, 1999.

[10] Singh L, Singh R, Zhang B, Kaushik B K, Kumar S. Localized Surface Plasmon Resonance Based Hetero-Core Optical Fiber Sensor Structure for the Detection of L-Cysteine. IEEE Transactions on Nanotechnology, 2020: 19, 201-208. 10.1109/TNANO.2020.2975297.

[11] Fang Y-L, Wang C-T and Chiang C-C. A small U-shaped bending-induced interference optical fiber sensor for the measurement of glucose solutions. Sensors, 2016: 16 1460. DOI: 10.3390/s16091460.

[12] Tian Y, Wang W, Wu N, Zou X, Wang X. Tapered optical fiber sensor for label-free detection of biomolecules. Sensors (Basel). 2011: 11 (4), 3780-3790. DOI: 10.3390/s110403780.

[13] Jarzebinska R, Korposh S, James S, Batty W, Tatam R and Lee S-W. Optical gas sensor fabrication based on porphyrin-anchored electrostatic self-assembly onto tapered optical fibres. Anal. Lett. 2012: 45, 1297–309. DOI: / 10.1080/00032719.2012.673097.

[14] Vahala K J. Optical microcavities. Nature, 2003: 424, 839–846. DOI: 10.1038/nature01939.

[15] Edwards P S, Janisch C T, He L, Zhu J, Yang L and Liu Z. Fibre taper based Raman spectroscopic sensing. Photonics Conf. (IPC) (IEEE), 2012, 501–2.

[16] Brambilla G. Optical fibre nanotaper sensors. Opt. Fibre Technol. 2010: 16 331–42. DOI: org/10.1016/j.yofte.2010.08.009

[17] Brambilla G. Optical fibre nanowires and microwires: a review. J. Opt., 2010: 12, 043001

[18] Lucas P, Coleman G J, Jiang S, Luo T and Yang Z. Chalcogenide glass fibers: optical window tailoring and suitability for bio-chemical sensing. Opt. Mater., 2015: 47, 530–536. DOI.10.1016/j.optmat.2015.06.034.

[19] Farnesi D et al. Quasi-distributed and wavelength selective addressing of optical micro-resonators based on long period fiber gratings. Opt. Express, 2015: 23, 21175–80. DOI.org/10.1364/OE.23.021175

[20] Lee B H, Kim Y H, Park K S, Eom J B, Kim M J, Rho B S and Choi H Y. Interferometric fiber optic sensors. Sensors, 2012: 12, 2467–2486.

[21] Huang, X., Li, X., Yang, J. et al. An in-line Mach-Zehnder interferometer using thin-core fiber for ammonia gas sensing with high sensitivity. Sci Rep., 2017: 7, 44994. DOI .org/10.1038/srep44994.

[22] Zhang W, Wang R, Rong Q, Qiao X, Guo T, Shao Z, Li J, Ma W. An optical fiber Fabry-Perot interferometric sensor based on functionalized diaphragm for ultrasound detection and imaging. IEEE Photonics, 2017: 9, DOI: 10.1109/JPHOT.2017.2694480.

[23] Kashyap R. Fiber Bragg Gratings, 2010. (San Diego, CA: Academic).

[24] Venkatesan, V. N., Ramalingam, R. Numerical and experimental investigation of FBG strain response at cryogenic temperatures. IOP Conf. Series: Materials Science and Engineering, 2017: 171, 012133. DOI: 10.1088/1757-899X/171/1/012133.

[25] Maguis S. et al. Biofunctionalized tilted fiber Bragg gratings for label free immune-sensing. Opt. Express, 2008: 16, 19049–62. DOI: .org/10.1364/OE.16.019049.

[26] Urrutia A., Goicoechea J., and Arregui, F. J. Optical Fiber Sensors Based on Nanoparticle-Embedded Coatings. Journal of Sensors, 2015: 2015, 805053. DOI: 10.1155/2015/805053.

[27] Seitz W R. Chemical sensors based on fibre optics. Anal. Chem., 1984: 56, 16–34. DOI: org/10.1021/ac00265a711.

[28] Wilhelm A, Lucas P, DeRosa D and Riley M. Biocompatibility of Te–As–Se glass fibers for cell-based bio-optic infrared sensors. J. Mater. Res., 2007: 22, 1098–1104. DOI: 10.1557/jmr.2007.0127.

[29] Anne M-L et al. Chalcogenide glass optical waveguides for infrared bio-sensing. Sensors, 2009: 9, 7398. DOI: 10.3390/s90907398

[30] Wiercigroch E. et al. Raman and infrared spectroscopy of carbohydrates: a review. Spectrochim. Acta A, 2017: 185, 317–335. DOI: 10.1016/j.saa.2017.05.045

[31] Korposh S, James S, Tatam R and Lee S-W. Fibre-optic chemical sensor approaches based on nanoassembled thin films: a challenge to future sensor technology. Optical Fiber, 2013 ed S W Harun (Rijeka: InTech) ch 9. DOI: 10.5772/53399.

[32] Turner A P F and Magan N. Electronic noses and disease diagnostics. Nat. Rev. Microbiol., 2004: 2, 160–166. DOI: 10.1038/nrmicro823

[33] Pavlou A K, Magan N, Jones J M, Brown J, Klatser P and Turner A P F. Detection of Mycobacterium tuberculosis (TB) in vitro and in situ using an electronic nose in combination with a neural network system. Biosens. Bioelectron., 2004: 20, 538–544. DOI: 10.1016/j.bios.2004.03.002.

[34] Haick H, Hakim M, Patrascua M, Levenberg C, Shehada N, Nakhoul F and Abassi Z. Sniffing chronic renal failure in rat models via an array of random network of singlewalled carbon nanotubes. ACS Nano, 2009: 3, 1258–1266. DOI: 10.1021/nn9001775

[35] Peng G et al. Diagnosing lung cancer in exhaled breath using gold nanoparticles. Nat. Nanotechnol., 2009: 4, 669–673. DOI: 10.1038/nnano.2009.235.

[36] Wolfbeis O S, Weis L J, Leiner M J P and Ziegler W E. Fiber-optic fluorosensor for oxygen and carbon dioxide. Anal. Chem., 1988: 60, 2028–2030. DOI: 10.1021/ac00170a009.

[37] Weiner I D, Mitch W E and Sands J M. Urea and ammonia metabolism and the control of renal nitrogen excretion. Clin. J. Am. Soc. Nephrol., 2014: 10, 1444–1458. DOI: 10.2215/CJN.10311013.

[38] Schmidt F M, Vaittinen O, Metsälä M, Lehto M, Forsblom C, Groop P H and Halonen L. Ammonia in breath and emitted from skin. J. Breath Res., 2013: 7, 017109. DOI: 10.1088/1752-7155/7/1/017109.

[39] Turner C, Španěl P and Smith D. A longitudinal study of ammonia, acetone and propanol in the exhaled breath of 30 subjects using selected ion flow tube mass spectrometry. SIFT-MS Physiol. Meas., 2006: 27, 321–337. DOI: 10.1088/0967-3334/27/4/001.

[40] Wolfbeis O S and Posch H E. Fibre-optic fluorescing sensor for ammonia. Anal. Chim. Acta, 1986: 185, 321–7. DOI: .org/10.1016/0003-2670(86)80060-5.

[41] Wang T, Yasukochi W, Korposh S, James S W, Tatam R P and Lee S-W. A long period grating optical fiber sensor with nano-assembled porphyrin layers for detecting ammonia gas. Sensors Actuators B, 2016: 228, 573–580. DOI: 10.1016/j.snb.2016.01.058.

[42] Rodríguez A J et al. A fiber optic ammonia sensor using a universal pH indicator. Sensors, 2014: 143, 4060–73. DOI: .org/10.3390/s140304060.

[43] Tiwari D, Mullaney K, Korposh S, James S W, Lee S-W and Tatam R P. An ammonia sensor based on Lossy mode esonances on a tapered optical fibre coated with porphyrinincorporated titanium dioxide. Sensors Actuators B, 2017: 242, 645–652. DOI: 10.1016/j.snb.2016.11.092.

[44] Shirasu M and Touhara K. The scent of disease: volatile organic compounds of the human body related to disease and disorder. J. Biochem., 2011: 150, 257–266. DOI: 10.1093/jb/mvr090

[45] Elosua C, Matias I R, Bariain C and Arregui F J. Volatile organic compound optical fiber sensors: a review, Sensors, 2006: 6,1440–1465.

[46] Hernandez F U et al. Optical fibre sensing during critical care. Proc. SPIE, 2017: 10340, 1034012.

[47] Schwalfenberg G K. The alkaline diet: Is there evidence that an alkaline pH die benefits health. J. Environ. Public Health, 2012: 2012, 727630. DOI: 10.1155/2012/727630

[48] Shao L-Y, Yin M-J, Tam H-Y and Albert J. Fiber optic pH sensor with self-assembled polymer multilayer nanocoatings. Sensors, 2013: 13, 1425–1434. DOI:10.3390/s130201425

[49] https://oceanopticscom/measurementtechnique/ph-sensing.

[50] https://presensde/products/ph/sensorshtml.

[51] Tiwari G et al. Drug delivery systems: an updated review. Int. J. Pharm. Invest., 2012: 2, 2–11. DOI: 10.4103/2230-973X.96920

[52] Korposh S, Chianella I, Guerreiro A, Caygill S, Piletsky S A, James S W and Tatam R P. Selective vancomycin detection using optical fibre long period gratings functionalised with molecularly imprinted polymer nanoparticles. Analyst, 2014: 139 2229–36. DOI: 10.1039/c3an02126b

[53] Li L, Ding H, Lia B D W and Chen J. Rapid detection of propofol in whole blood using an automated on-line molecularly imprinted pretreatment coupled with optical fibre detection. Analyst, 2012: 137, 5632. DOI: 10.1039/c2an35523j

[54] Zhu, G., Singh, L., Wang, Y. et al. Tapered Optical Fiber-Based LSPR Biosensor for Ascorbic Acid Detection.

Photonic Sens.: 2020). DOI: 10.1007/s13320-020-0605-2.

[55] Urrutia A., Goicoechea J., and Arregui, F. J., Optical Fiber Sensors Based on Nanoparticle-Embedded Coatings. Journal of Sensors, 2015: 2015, 805053. DOI: 10.1155/2015/805053.

Chapter 8

Photonics for AI and AI for Photonics: Material and Characteristics Integration

Sunil Sharma and Lokesh Tharani

Abstract

We are living in the technological era, where everything is integrated with each other. If we are discussing regarding communication, it is integrated with one or two technologies. If we are discussing regarding automation, discussing regarding Image processing, discussing regarding embedded system, they all are integrated with a combination of technologies. Correspondingly Artificial Intelligence (AI) and Photonics are also integrated with each other. Now a day as AI is utilizing with photonics in abundant fields as well photonics is also serving AI to facilitate ultrafast AI networks to offer a novel class of Information Processing Machines (IPM). This chapter is based on identification and implementation of photonics for AI utility and AI for photonics. In this category a Dual core Photonics crystal fiber (PCF) is proposed which serve to identify infected cells of human being along with the integration of AI. This proposed design of PCF is providing relative sensitivity and confinement loss in an optimized manner with the impact of AI. Here potency of AI as well as of Photonics is explained to serve their applications related to each other.

Keywords: Artificial Intelligence, Fiber Optics and Photonics, Optical Networks, Photonic Crystal Fiber, Integration

1. Introduction

Latest technological development in photonics has multiplied only due to integration of photonic platform with conception of Opto-electronic elements [1]. The Photonic Integrated Circuits (PICs) [2] have facilitated the ultrafast Artificial Neural Networks (ANN) [3], to propose a novel class of Information Processing Machines (IPM) [4]. There are number of reasons available which reveals that photonics is somewhere associated with AI. In this direction the latest example can be considered as development of Neuro-morphic [5] electronics, which shows that *high processor delay* can be eliminated by offering a consequent technology to extend the vicinity of AI. It offer sub-nanosecond [5] delay and consequently conquers challenges in terms of present and future aspects.

This latest developed technology 'Neuro-morphic electronics system' [5] is integrated with most recognized technology which is known as semiconductor photonics. It is composed of third and fifth group of elements i.e. GaAs and InP [2].

IntechOpen

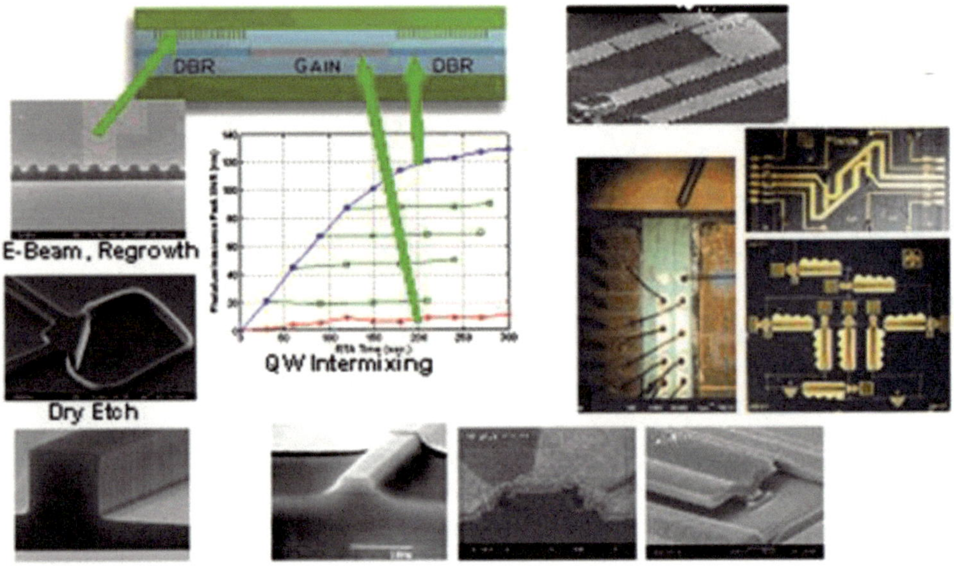

Figure 1.
GaAs & InP Composed Photonic Integrated Circuits [2].

Below **Figure 1** represents the photonic integrated technology indicating fabrication, characteristics like growing and mixing of GaAs and InP materials to provide efficient, robust, and monolithic optoelectronic integration platform. It was developed and observed by Sandia National laboratory services.

The developmental growth of photonic crystals, components and meta-materials [6] lead to the advancement of photonics in the area of designing, modeling and technological integration. This kind of integration investigates AI with photonics. This promising domain is someway sustained by 'photonic materials' [7] which assist to find out and intend innovative applications of AI. It should be noted down that how photonics is contributing for the implementation of AI tools and techniques.

The contributing field of photonics towards AI includes Neuro-morphic electronic system, Optical Neural Network (ONN), Nano Photonics,

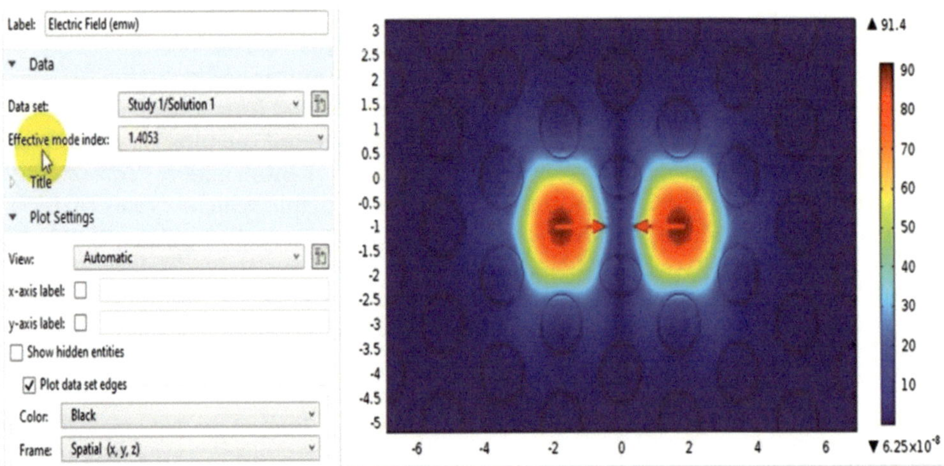

Figure 2.
Proposed dual cores PCF with different mode indexes.

meta-materials, optical sensing, optical imaging [8], optical computing, Information Processing Machines etc. These above mentioned optics emerging domains can be integrated with AI tools [9] to enhance the efficiency and performance of these systems.

Figure 2 represents the design structure of dual core silica PCF with an effective index mode of 1.4053. By changing the mode index value we can have light confinement variation which is shown below in **Figure 3 (a & b)**.

As shown below in **Figure 4** indicates contribution of photonics in terms of machine intelligence with Neuro-morphic computing along with Optical neural network and optical sensing for AI technology. These latest technologies helped AI to diagnose critical disease.

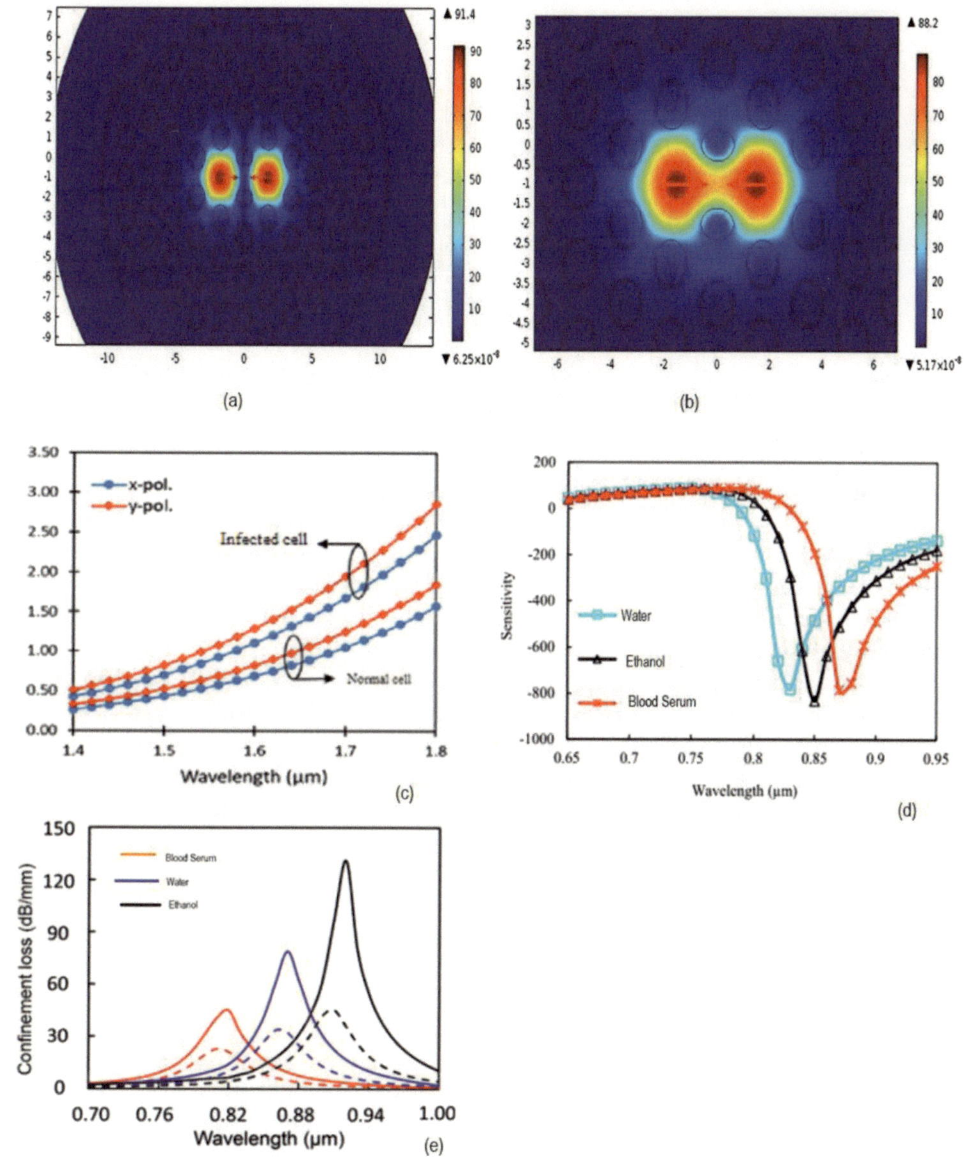

Figure 3.
(a, b) Light Confinement through proposed design for different values of Index Modes (c) Identification of infected cells with AI (d) Relative sensitivity (e) confinement loss of proposed design.

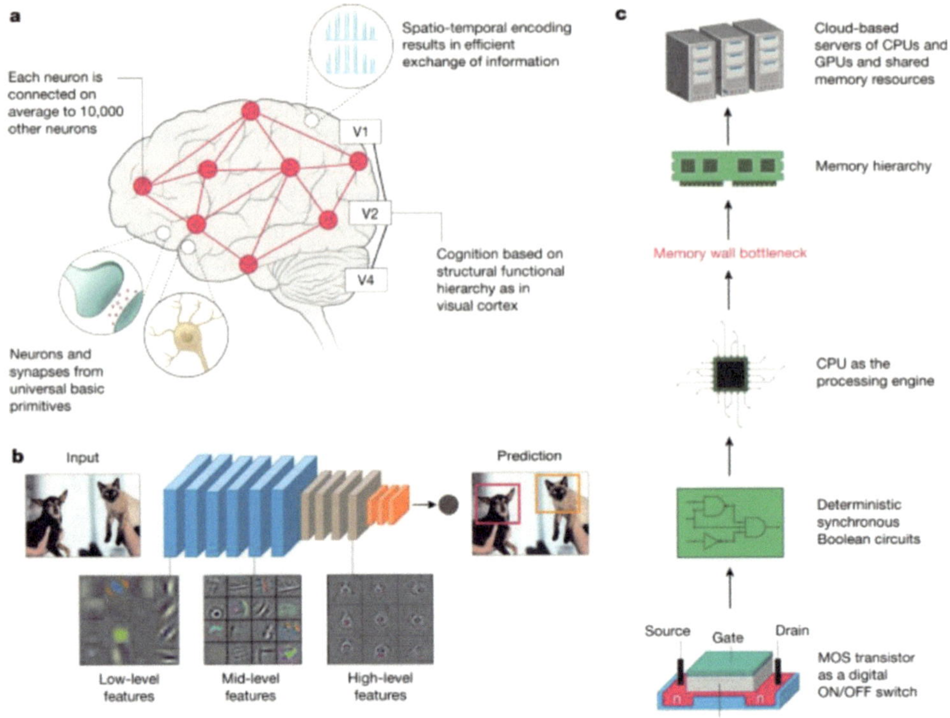

Figure 4.
Contributing field of photonics for AI (a) AI with Neuro-morphic computing (b) Optical Neural Network (c) Optical sensing and computing [5].

2. Photonics materials and their characteristics for AI

We all are witnessing an inconceivable age of drastically development in applications that necessitate expansion in AI [10]. If we are discussing about the ingenious novel outcomes that are gradually trending towards the market place and many more are preferred and expected. Fiber Optics & Photonic materials [11] are widely used for these products like new display, personalized mobile devices, novel sensors, and new information processing machining products for both storage and data processing. It is trending in very clear manner that the areas of Fiber Optics & Photonic materials are fundamental technologies for the globe. Inventing and uncovering new materials [12] in the Fiber Optics & Photonics domain will be exceedingly critical to see more and more novel outcomes to improve normal people's lives.

Materials that have been exposed at the crucial point of life, always changes the history of human being along with the country. Materials that are used to senses, materials that are used to stores, materials that can be used as energy efficient, some translucent materials which can be folded easily and some materials that are manufacturable at low cost. New discovered materials such as doped silica materials [13], resistance changing materials and spontaneously magnetize and polarize materials have been discovered and using widely for AI integration and their applications.

In the line of discovery of new materials, the Picometer [7] can also be considered as a vibrant example in the field of atomic structures. There are numerous atomic structures available that were simulated and their data were utilized for AI analysis to identify artificially controlled 'oxygen octahedral rotation' (OOR) patterns as shown in below **Figure 5**.

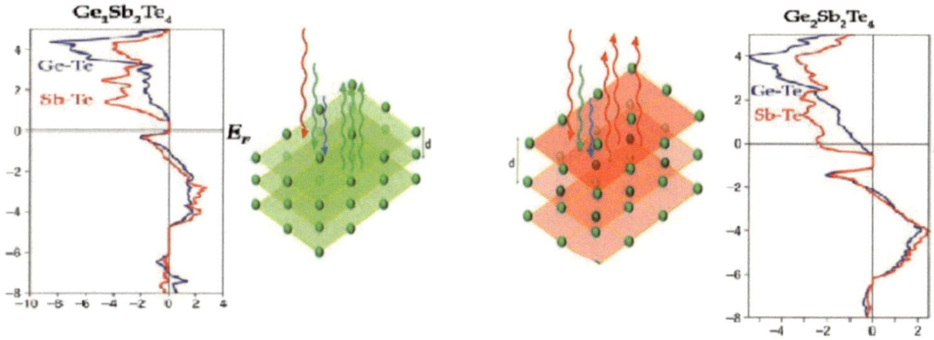

Figure 5.
Oxygen Octahedral Rotations and its characteristics [7].

It was used as Disorder-Driven Metal–Insulator Transition in Crystalline Vacancy-Rich Ge-Sb-Te Phase-Change Materials [7].

2.1 New Investigative Materials for AI

The discovery and development in the new materials plays an important role in the technological progress. As we have already seen that how silica has revolutionized the microelectronics industry. Materials discovery and design efforts require interplay between materials prediction, synthesis and characterization [12] have increased applications of computational tools and techniques, increased generation of material's databases, and accelerated advances in experimental methods significantly. Some of them are composed of three special elements i.e. germanium, antimony and tellurium which is defined as Ge-Sb-Te alloy [2] and can be termed as phase-change memory materials. This alloy is selected from the group of chalcogenide glass (As_2Se_3) [12] which can be used in rewritable optical discs.

The above mentioned **Figure 6** is used as a non-volatile quasi-continuously reprogrammable platform. This phase-change memory material rapidly changes its atomic structure from crystalline to solid amorphous when swiftly melted in presence of temperature. These kinds of materials are widely used in 'electronic memory' applications of AI tools such as *data storage*. Even though there are countless integration is possible with Ge-Sb-Te alloy, the new material **GST467** [6] revealed by CAMEO (Closed-Loop Autonomous System for Materials Exploration and Optimization) is most favorable for phase-changing applications.

CAMEO found the best Ge-Sb-Te alloy that had the largest difference in "optical contrast" [6]. GST467 also found applications in photonic switching devices that can be used to control the direction of light in given circuit. These devices can also be utilized in Neuro-morphic computing [5], which is an emerging field focusing on development of devices which imitate the formation and role of neurons in human brain. Materials science or solid-state physics is plagued by the 'curse of dimensionality'.

3. AI for photonics

When the words "artificial intelligence" (AI) comes to mind, our first thoughts may be of super-smart computers or robots that perform tasks without needing any help from humans.

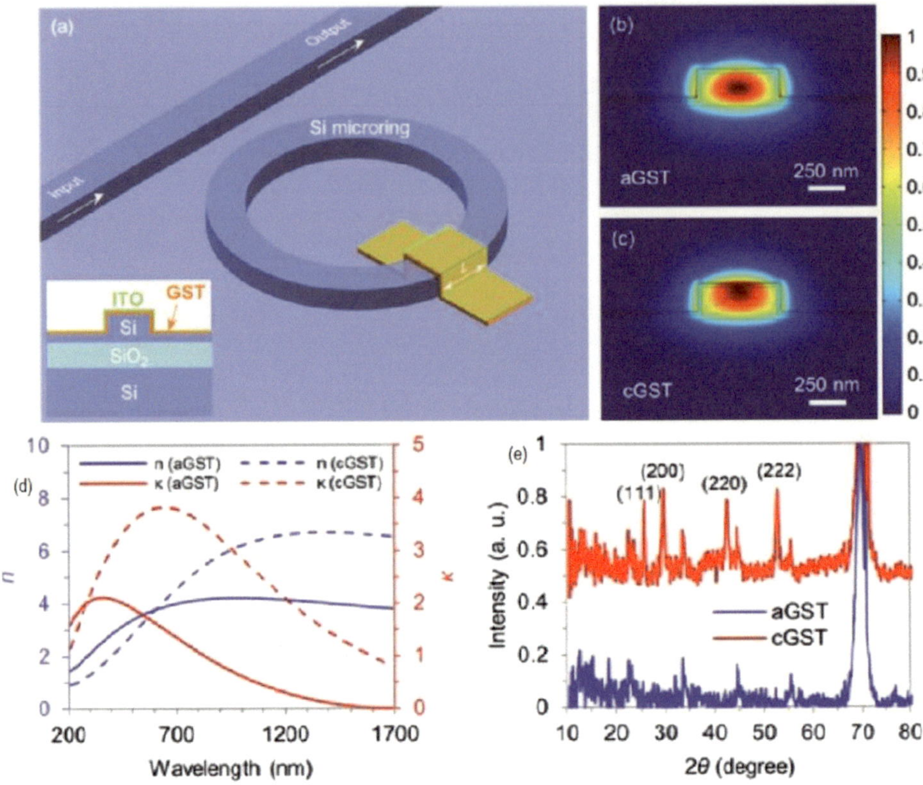

Figure 6.
GST467 with AI (a) Schematic cross-section of the hybrid waveguide. (b)&(c) Fundamental quasi-transversal electric (TE) mode profiles of the hybrid waveguide at 1550 nm for (d) complex refractive index of GST and GST as a function of wavelength. (e) XRD data of GST [6].

A multi-institutional team of research scholars from National Institute of Standards and Technology (NIST) [6] have developed an AI algorithm known as CAMEO. It was used for the discovery of potentially applicable new photonic material without any additional preparation and efforts from the scientist. These AI systems helped to reduce the trial-and-error time which generally scientists use up in the lab. Along with this these systems maximizes the productivity and efficiency of their research work. Another research scientists team at POSTECH (Pohang University of Science and Technology) [7] got succeed in creating a novel substance that generates electricity by effect of polarization at room temperature. The variation so observed would be confirmed in crystal structure by analysis of deep neural network. The above mentioned examples revealed the techniques behind making materials used in new memory devices by using artificial intelligence. So it is very much clear that the use of modern computational techniques like AI can be used to improve the rate of discovery of these new photonics materials and vice versa. Helping scientists in reaching their outcomes more efficiently and quickly by performing only few experiments with limited resources. All these things became possible only because of integration of AI and Photonics.

The optical properties are typically calculated by using Maxwell's Equations [13]. The desired optical response can be obtained by adjusting the initial design and performing multiple simulations until the outcome is achieved. Despite designing issues AI can help optics and nano photonics in different tasks, for example AI used to estimate the optical properties of black carbon fractal aggregates. Another example is reported where they combines finite element simulations and clustering for the identification of photonic modes [14] with large local field energies and specific spatial properties. It is shown that the combination of machine learning with photonics [15] can revolutionize one of the most important fields in optical imaging.

4. Proposed dual core PCF design integrated with AI

The silica glass is easily available and have some characteristics due to this it is preferred for designing PCF structures. Below **Table 1** depicts some properties of silica glass [15].

Silica is the purest form of SiO_2 which is easily available from the sand as a raw material. This raw silica is used to convert into Electronic Grade Silicon (EGS) from various processes. This glass has superior transmission chatcteristics in the UV (Ultra-violet) and IR (Infra-red) spectra, a very low dielectric coefficient and excellent properties where fluorescence or polarization is an issue. This silica can be shaped too many forms and sizes. It has excellent resistance to non-fluorinated acids, solvents and plasmas. The finite-difference method is the most accurately and numerically efficient method to solve Maxwell's Equation [15] and needs less computational time.

By selection of Silica glass as a core material for designing of PCF structure, below mentioned **Figure 7** depicts the cross-sectional view of proposed dual core Silica PCF with circular sensing ring. The diameter of the air hole is 1.2 μm. Here elliptical air hole is also used in the first layer and the semi major and semi minor axis for that ellipse is 1.2 and 0.8 μm respectively. The pitch value for the proposed structure is 2 μm.

After designing the structure of Dual Core PCF if there is a variation of index mode then due to different mode index values, there must be some variation measured in confining light through designed PCF. This variation is already mentioned in above **Figure 3 (a & b)**.

It indicates that as index mode value varies like 1.4053, 1.4055, 1.4088, 1.41 … .. The variation is observed in confining the light through core of the proposed fiber.

The proposed dual core PCF for sensing various applications like blood sample detection, alcohol detection, disease detection, White Blood Cells (WBC), Red Blood Cells (RBC) detection and for many more pathological detection can be integrated with AI technology which provides optimized results to diagnose infected cells. For this purpose below mentioned setup as shown in **Figure 8** is

Properties	Silica Glass
Density (g/cm^3)	2.2
Refractive Index (micrometer)	1.458
Light Transmission wavelength (micrometer)	0.18–2.5
Max Temperature (Degree Centigrade)	1120
Poission's Ratio	0.17
Specific heat capacity (J/Kg-K)	720
Speed of sound (m/s)	180×10^3

Table 1.
Properties of Silica Glass Material [12].

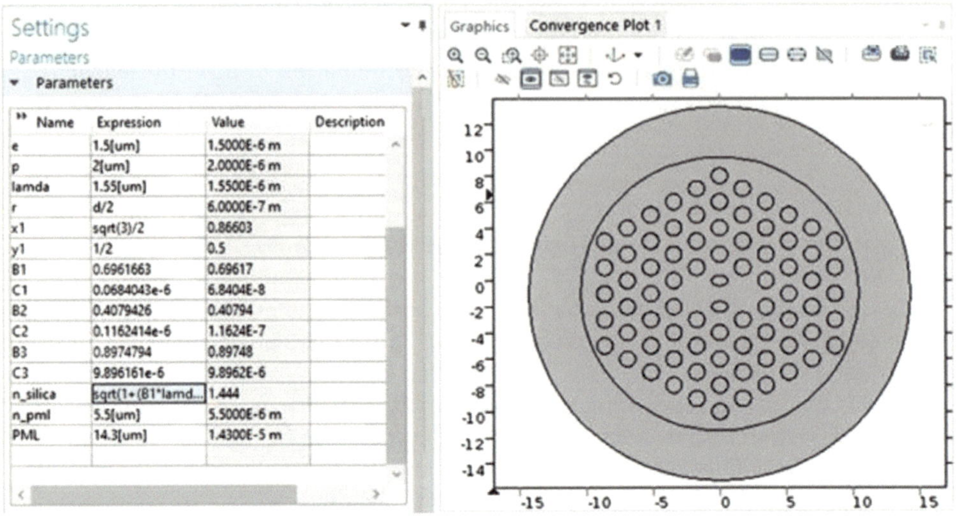

Figure 7.
Proposed Dual Core PCF with Perfectly Matched Layer (PML) Boundary.

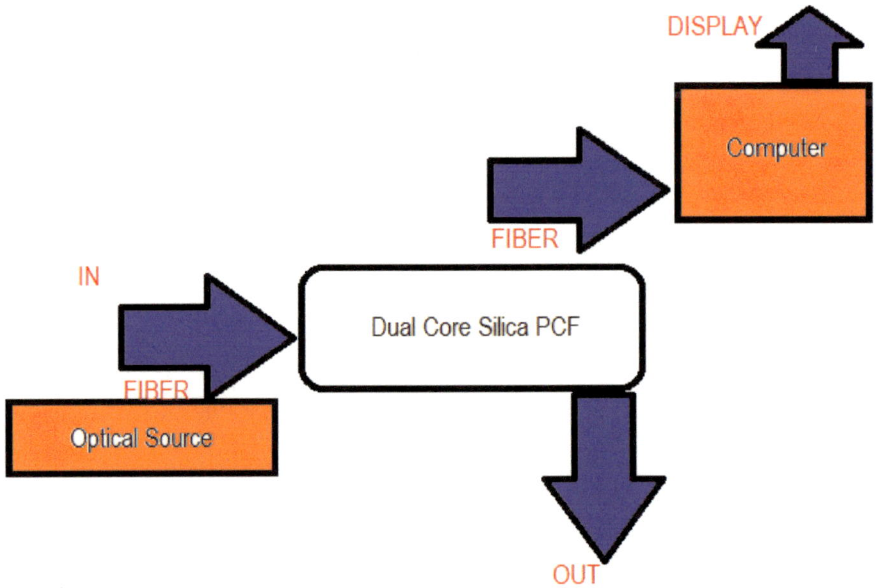

Figure 8.
Experimental Setup to Obtain Outcomes.

arranged. With the help of this setup the proposed dual core PCF can be utilized with AI to serve better and improved outcomes. The above mentioned **Figure 3 (c)** represents the infected cell by using designed PCF structure integrated with AI. Relative sensitivity and confinement loss is also displayed in **Figure 3 (d & e)**.

In this setup the optical source is used to supply power to the Fiber. By using the splicing technique fiber can be connected with the proposed PCF. IN and OUT ports are used to control the unknown analytes whose refractive Index (RI) need to be identified. When analyte interacts then the variations in terms of lows and peaks occurs which can be observed and displayed using computer. The outcomes so obtained can be enhanced to provide efficient result with AI. Dual core silica PCF

Parameter Tested	Refractive Index	Relative Sensitivity (%)	Confinement Loss (dB/km)
Ethanol	1.33	56.90	2.37×10^{-6}
Blood Serum	1.39	46.51	3.814×10^{-10}
Water	1.32	53.57	8.063×10^{-11}

Table 2.
Test Performed for various parameters.

serve as a sensing element used to sense the selected parameter and the AI technology boost the effects of results so obtained.

With the Above mentioned proposed design the following tested have been performed using Dual core Silica PCF.

Depending upon the refractive index of blood serum, the intensity of light is modulated and detected at other end of PCF [16]. The relation between evanescent field absorbed by sensing species and intensity modulation at output end is observed.

Sensitivity is obtained by using

$$r_f = f\left(\frac{n_r}{n_c}\right) \tag{1}$$

Where n_r is the refractive index of the fluid, n_c is core refractive index, r_f is relative sensitivity coefficient and 'f' is the ratio of optical power with in large holes to the total power which is given as

$$f = \int\left[\left(E_x H_y - H_x E_y\right)_{samples} / \int \left[\left(E_x H_y - H_x E_y\right)\right]_{total}\right. \tag{2}$$

Confinement loss [17, 18] is calculated by

$$L_C = (40\pi / \ln\ (10)\ \lambda)\ \mathrm{Im}\ (\mathrm{neff})\ [\mathrm{dB/km}] \tag{3}$$

or it can be written as

$$L_C\ (\mathrm{dB/m}) = 8.686\, k_0\, \mathrm{I_m}\left(n_{eff}\right) \times 10^6 \tag{4}$$

Here n_{eff} signifies imaginary part of effective refractive index, and k_0 is the free-space number.

The data set of blood serum, ethanol and water for this case of investigation is selected as an input which can be passed through the setup and results so obtained have been optimized by using AI. These results obtained numerically and experimentally have been presented in above mentioned **Table 2**.

5. Discussion

The potency of the AI standards lies in its capacity to deal with anonymous computing troubles. It is practically identified that it is giving not only innovative or optimized solutions and forecasting, but also original substantial impending to the structure by using integration with technologies. Here we have presented an integrated discussion between AI and Photonics. The AI has been utilized to nurture tiny investigational datasets in iterative method to envisage new materials and execute multi objective optimization of properties for selected materials. Correspondingly Photonics is also offering new materials for booming realization and performing computation takes in an efficient manner to AI. The characteristics of the dual-core photonic crystal fiber (PCF) sensor are studied using the finite element method (FEM), and the structure is improved according to the numerical simulation results.

6. Conclusions

In the revolutionary field of optics and photonics, most of the work has so far been offered on purpose of photonics to the realization of AI to the intend, expansion and optimization of photonic meta-materials and various devices. AI techniques present prospects both to expand physical approaching and to investigate constraints in a more proficient manner.

Most successful paradigms of AI and photonics like Neuro-morphic electronic system, Optical Neural Network (ONN), Nano Photonics, meta-materials, optical sensing, optical imaging have also been demonstrated here in this chapter in which AI is boosting photonics and similarly photonics is also helping AI to perform efficiently. The proposed Dual core Silica PCF is used to identify infected cell in a human body. Due to easily presence of Silica glass and its vibrant characteristics it is preferred for the proposed PCF design. The refractive index of selected material is 1.458, Specific heat capacity is 720 J/Kg-K, Light Transmission wavelength is 0.18–2.5micrometer. It has been observed that the relative sensitivity for ethanol, blood serum and water is 56.90%, 46.51% and 53.57% respectively. Similarly the confinement loss for the proposed structure is 2.37×10^{-6}, 3.814×10^{-10} and 8.063×10^{-11} dB/km respectively for the same parameters as mentioned above.

Acknowledgements

I would like to acknowledge each and everyone those who have helped me directly or indirectly to complete this research article. I tried to cite all the resources at my best, but if i forgot someone then kindly receive my apologies in advance.

Conflict of interest

I declare no conflict of interest for this research article.

Data availability

The data that support the findings of this study are available from the corresponding author upon reasonable request.

DOI: http://dx.doi.org/10.5772/intechopen.97781

Author details

Sunil Sharma* and Lokesh Tharani
Department of Electronics Engineering, Rajasthan Technical University, Kota, India

*Address all correspondence to: ersharma.sunil@gmail.com

References

[1] Piccinotti D., MacDonald K.F., Gregory S.A.,, Youngs I. and Zheludev N.I. (2020) Artificial intelligence for photonics and photonic materials, IOP Publishing Ltd. Rep. Prog. Phys. 84 012401

[2] Wang, J. J., Xu, Y. Z., Mazzarello, R., Wuttig, M., & Zhang, W. (2017). A Review on Disorder-Driven Metal-Insulator Transition in Crystalline Vacancy-Rich GeSbTe Phase-Change Materials. Materials (Basel, Switzerland), 10(8), 862. https://doi.org/10.3390/ma10080862

[3] Soler M, Estevez M.C., Rubio M.C., Astua A., and Lechuga L.M., (2020) How Nano photonic Label-Free Biosensors Can Contribute to Rapid and Massive Diagnostics of Respiratory Virus Infections: COVID-19 Case A CS Sensors, 5 (9), 2663–2678 doi: 10.1021/acssensors.0c01180.

[4] V Goda K., Jalali B., Lei C., Situ G., and Westbrook P., (2020) AI boosts photonics and vice versa, APL Photon. 5, 070401, doi: 10.1063/5.0017902.

[5] Roy, K., Jaiswal, A. & Panda, P., (2019) Towards spike-based machine intelligence with Neur-omorphic computing. Nature 575, 607–617. https://doi.org/10.1038/s41586-019-1677-2.

[6] Zeng J., Khanolkar A., Xu P., Colburn S., Deshmukh S., Myers J., Frantz J., Pop E., Hendrickson J., Doylend J., Boechler N., and Majumdar A., (2018) GST-on-silicon hybrid nanophotonic integrated circuits: a non-volatile quasi-continuously reprogrammable platform," Opt. Mater. Express 8, 1551–1561

[7] Peng Chen, Mathieu N. Grisolia, Hong Jian Zhao, Otto E. González-Vázquez, L. Bellaiche, Manuel Bibes, Bang-Gui Liu, and Jorge Íñiguez, (2018) Energetics of oxygen-octahedra rotations in perovskite oxides from first principles, Phys. Rev. B 97, 024113

[8] Wei, J.; Yi, L.; Giacoumidis, E.; Cheng, Q.; Tao Lau, A.P., (2020) Special Issue on "Optics for AI and AI for Optics, Appl. Sci. 10, no. 9: 3262. https://doi.org/10.3390/app10093262

[9] Soler M., Scholtz A., Zeto R., and Armani A.M., (2020) Engineering photonics solutions for COVID-19, APL Photonics 5, 090901, https://doi.org/10.1063/5.0021270

[10] Taha, B.A.; Al Mashhadany, Y.; Hafiz Mokhtar, M.H.; Dzulkefly Bin Zan, M.S.; Arsad, N., (2020) An Analysis Review of Detection Corona virus Disease 2019 (COVID-19) Based on Biosensor Application, Sensors, 20, 6764. https://doi.org/10.3390/s20236764

[11] Yao K., Unni R. and Zheng Y., (2019) Intelligent nanophotonics: merging photonics and artificial intelligence at the nanoscale, De Gruyter | 2019 doi: https://doi.org/10.1515/nanoph-2018-0183

[12] Sharma R.K., S. Sharma and Vyas K., (2018) Analysis of Different Types of Core Materials in Photonic Crystal Fiber, 5th IEEE Uttar Pradesh Section International Conference on Electrical, Electronics and Computer Engineering (UPCON), Gorakhpur, India, pp. 1–6, doi: 10.1109/UPCON.2018.8597132

[13] Sharma S., Sharma R.K., Gupta R., Dash P. (2020) Design and Analysis of Elliptical Core Spiral Silica Photonic Crystal Fiber with Improved Optical Characteristics. In: Kumar A., Mozar S. (eds) ICCCE 2019. Lecture Notes in Electrical Engineering, vol 570. Springer, Singapore. https://doi.org/10.1007/978-981-13-8715-9_9

[14] Jain A., Sharma R.K., Agarwal V., Sharma S. (2020) A New Design of Equiangular Circular Cum Elliptical Honeycomb Photonic Crystal Fiber. In: Ranganathan G., Chen J., Rocha Á. (eds) Inventive Communication and Computational Technologies. Lecture Notes in Networks and Systems, vol 89. Springer, Singapore. https://doi.org/10.1007/978-981-15-0146-3_6

[15] Sharma S., Tharani L., Sharma R.K. (2020) Designing a Nonlinear Tri Core Photonic Crystal Fiber for Minimizing Dispersion and Analyzing it in Various Sensing Applications. In: Mathur G., Sharma H., Bundele M., Dey N., Paprzycki M. (eds) International Conference on Artificial Intelligence: Advances and Applications 2019. Algorithms for Intelligent Systems. Springer, Singapore. https://doi.org/10.1007/978-981-15-1059-5_3

[16] H. Ademgil, Highly sensitive octagonal photonic crystal fiber based sensor, Optik-Int. J. Light Electron Opt. 125 (20) (2014) 6274–6278

[17] V. Kaur and S. Singh, "Performance analysis of multichannel surface plasmon resonance sensor with dual coating of conducting metal oxide," J. Nanophotonics 12(1), 016012 (2018).

[18] U. S. Dinish et al., "Highly sensitive SERS detection of cancer proteins in low sample volume using hollow core photonic crystal fiber," Biosens. Bioelectron. 33(1), 293–298 (2012).

Chapter 9

Laser Opto-Electronic Oscillator and the Modulation of a Laser Emission

Alexander Bortsov

Abstract

The autonomous optoelectronic generator (OEO) is considered in the chapter as a source of low-noise oscillations. Differential equations are considered and methods with OEO modulation with direct and external modulation are analyzed. The complexity of both approaches is related to the non-standard way of description of the nonlinear method modulation for the internal (direct) structure and the utilization of the specific Mach-Zehnder modulator for the first stage on external modulation. The purpose of the presentation is to consider the main features of OEO as a low-noise generator. This includes consideration based on the study of differential equations, the study of transients in OEO, and the calculation of phase noise. It is shown that different types of fibers with low losses at small bending radii can be used as a FOLD in OEO. The important role of the choice of a coherent laser for OEO with a small spectral line width is shown. The prospects of using structured fibers with low losses at bends of less than 10 mm in OEO are described. The results of modeling dynamic processes in OEO with direct modulation are presented.

Keywords: opto-electronic oscillator, phase noise, optical fiber, QW laser, microwave oscillator

1. Introduction. The opto-electronic oscillator structure

Development and creation of the compact ultra-low-noise microwave signal sources, which would be impact-resistant, is an important problem of modern radio-physics and radio engineering. Levels of the phase noise spectral density at the microwave source output must be for most of the applications −120 ... -170 dB/Hz at generation frequency 8 ... 12 GHz for 1-kHz offset from a carrier. Constructions of these oscillators must sustain the strong mechanical impact loads in 200 ... 2000 N/cm and high accelerations up to 2 ... 10 g. Geometrical dimensions of the modern signal sources should often be approximately 10x10x10 cubic mm, especially for the satellite applications.

Development and implementation of new compact microwave and millimeter-wave oscillators with improved performance would lead to revolutionary jump in radio electronics, perhaps, comparable to discovery of the quantum-dimensional lasers or (as in radio engineering) at arriving of the high-stability quartz crystal resonator. The new type of oscillators called as opto-electronic oscillator (OEO) described in this paper will permit to use in the mobile communications and in

Internet systems of new radiofrequency channels for information transmission, including 30 ... 75-GHz ranges at the low power of transmitters. A number of publications devoted to OEO experimental investigations grows each year [1–8].

Opto-electronic oscillators will undoubtedly find wide application in the fiber-optical communication lines as well as in on-board radar systems on millimeter- and centimeter ranges, in communication systems as low-noise local oscillators in receivers and as a master clock in transmitters, in an optical lidar technology, as sensors of different physical quantities and in many other systems [8–16].

OEO diagrams with the direct modulation (OEO DM) presented in **Figure 1a** and the OEO structural diagram with *external modulation* of optical emission, which is often called as an opto-electronic oscillator with the Mach–Zehnder modulator (OEO MZ) presented in **Figure 1b.**

Let us consider the case of OEO operation with a small modulation index, and under the condition that the width of the spectral line of the optical laser

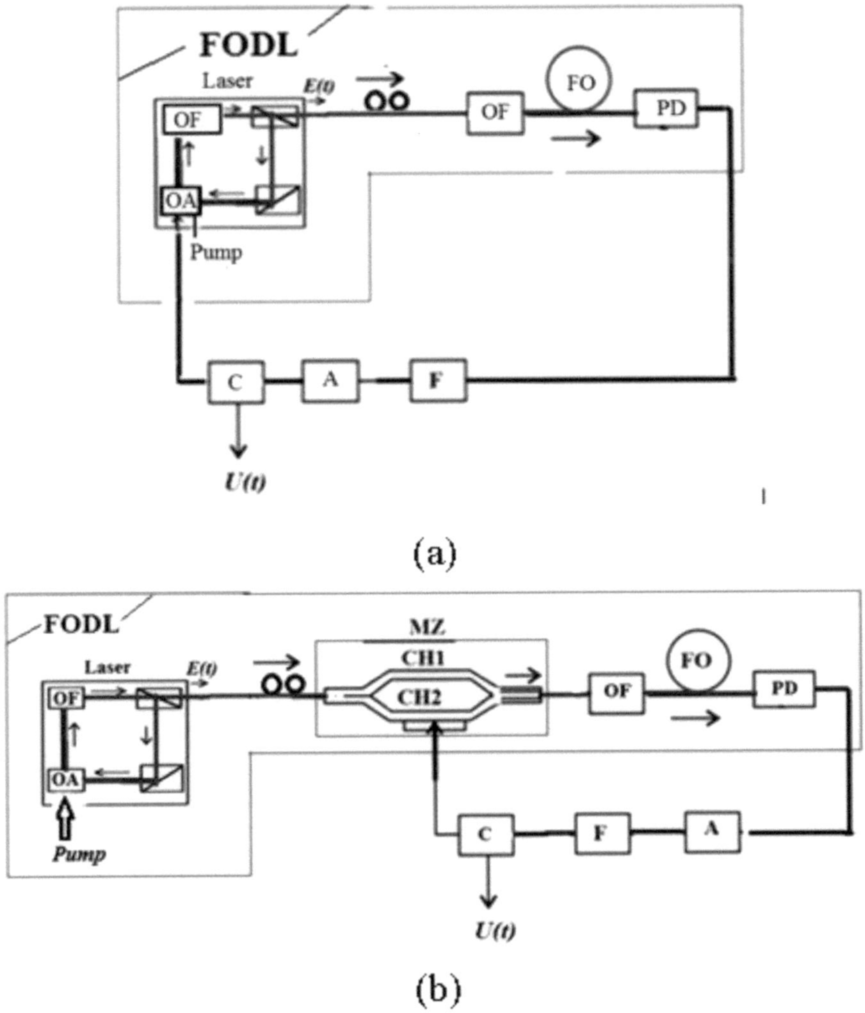

Figure 1.
Structural diagrams of the optoelectronic oscillator (a) with the direct modulation by current and (b) with external Mach-Zehnder modulator. Laser = the optical quantum generator (the laser or QWLD), MZ = the electro-optical MZ modulator, OA = the optical amplifier, OF = the optical filter, OF = the optical fiber, P_p = the pumping power, PD = the photo-detector, NA = the nonlinear amplifier, F = the RF filter, C = the RF coupler, CH1, CH2 = the optical channels of the MZ modulator.

generation$\Delta\nu_L$ is much smaller than the radio frequency f_0 of the OEO generation: $\Delta\nu_L << f_0$. In this case, the spectrum of the modulated radiation can be represented by several harmonics. We will limit our analysis to the case of two or three optical harmonics, respectively, with frequencies $\nu_1 = \nu_1 - f_0$, $\nu_2 = \nu_0$, $\nu_3 = \nu_3 - f_0$. Two of these optical frequencies ν_1 and ν_2 are spaced from the central optical laser frequency ν_0 by the sub-carrier frequency f_0.

Figure 2a and **b** shows the analog model of statistical processes in OEO MZ with utilization of the random variables correlator. The correlator structure is described in [**16**]. It consists of the multiplier "* ", two optical channels with different delays, and the delay cell defining by he delay in the optical fiber. The functional diagram **Figure 2a** and **b** illustrating principles of the correlator method and the frequency

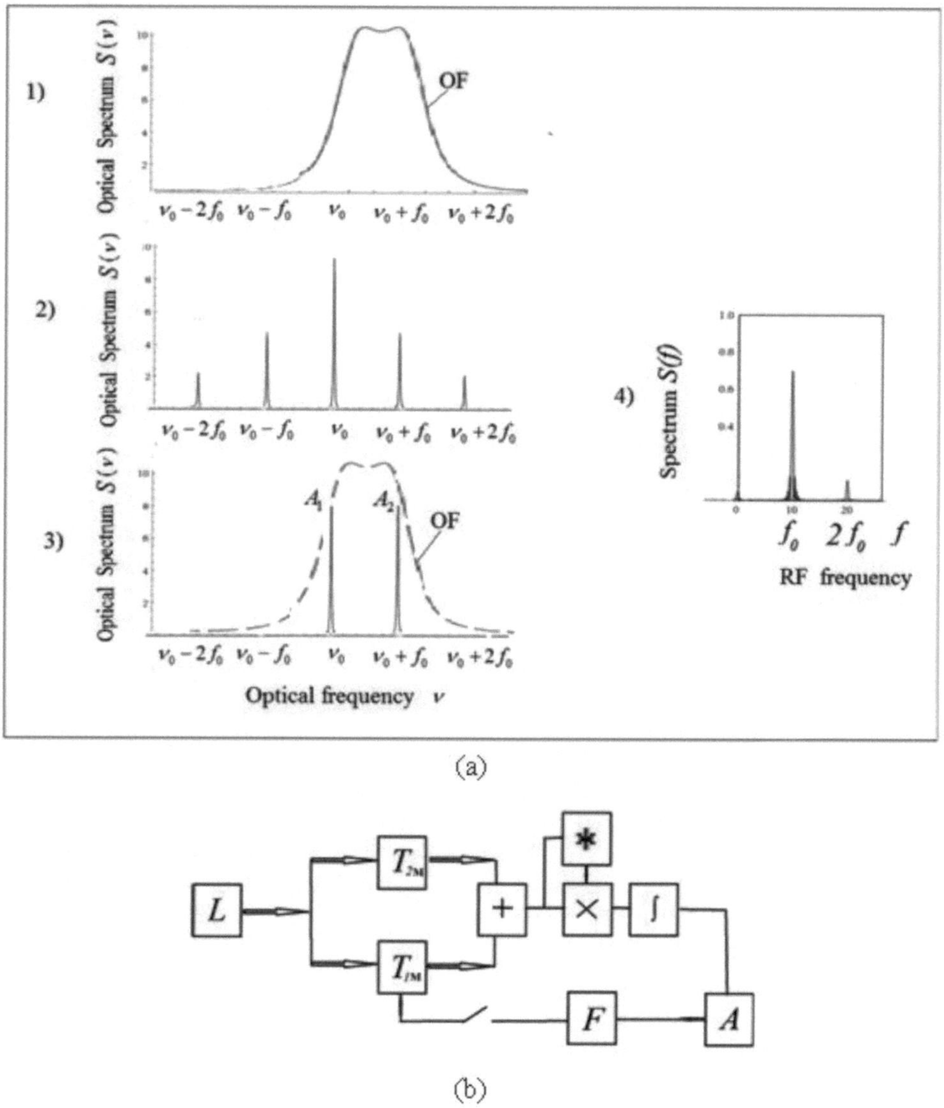

Figure 2.
The functional diagram illustrating principles of the correlator method and the frequency discriminator method in OEO with the MZ modulator and in the circuit with direct amplitude modulation at suppression of the one harmonic. a) Diagrams (1–4) of optical frequency selection; b) L = the laser, T_{1M} *and* T_{2M}*, = delay lines have delay times in channels, " + " = (adding), "x" = (multiplication), "*" = (conjugate operation), " ∫ " = (integration), F = the low-pass filter, a = the amplifier.*

discriminator method in OEO with the MZ modulator and in the circuit with direct amplitude modulation at suppression of the one harmonic.

2. OEO with direct modulation and OEO with the Mach Zehnder modulator

As in [8–16] when studying noise, OEO is considered here as an optoelectronic system in which oscillations are formed in the optical and radio frequency ranges. The oscillation frequency of the laser is approximately 200 Hz, and the radio frequency of the OAO generation is approximately 10 GHz. A VLD quantum-dimensional laser diode is used to generate laser radiation. The positive feedback ring is formed by an optoelectronic circuit consisting of a modulator, an optical fiber, a photodetector, an electronic amplifier, an electronic filter, and a directional coupler.

Fluctuations are formed at the OEO output. Laser fluctuations are of a quantum nature. When using a low-noise amplifier, the phase fluctuations of the laser determine mainly the noise of the OEO output.

3. OEO construction and its operation principle

Figure 1a shows a direct modulated OEO diagram. At the same time, the QWLD laser it works in the mode of amplitude modulation or intensity modulation. In **Figure 1b** the diagram of the OEO with external modulation is presented.

The OEO DM diagram (**Figure 1a**) is formed by a QWLD laser; a single-mode optical fiber (FO); a photodetector (FD); an electronic amplifier (A); an electronic filter (F), for example, based on a dielectric microwave resonator. The OOO MZ diagram (**Figure 1b**), in addition to the QWLD laser, includes elements that form a closed loop: the Mach Zehnder modulator (MZ); a fiber-optic system (FOS) containing an optical filter (OF) and single-mode optical fiber (FO); photodetector (PD), such as a quantum photodiode size; narrow-band RF filter (F), nonlinear amplifier (A), and directional coupler (C). A fiber-optic delay line (RF FODL) is formed by a laser connected in series, OF, FO, and PD (**Figure 1a**), or by a laser connected in series, MZ, OF, FO, and PD (**Figure 1b**). OEO can be considered as a delayed feedback oscillator.

Figure 1a and **b** show a laser as a source of optical oscillations, which includes a closed-loop optical amplifier (OA) and an optical filter (OF). We consider the case of the laser radiation modulation mode for single-mode, single-frequency, and linearly polarized optical radiation. When self-excitation conditions are met in such OEO systems (**Figure 1**) generation of microwave range oscillations occurs. A fiber-optic delay line (RF FODL) is formed by a laser connected in series, OF, FO, and PD (**Figure 1a**), or by a laser connected in series, MZ, OF, FO, and PD (**Figure 1b**). OEO can be considered as a delayed feedback oscillator.

Figure 1a and **b** shows a laser as a source of optical oscillatons, which includes a closed-loop optical amplifier (OA) and an optical filter (OF). The pump or pumping power P_p of the laser is shown conditionally.

We consider the case of the laser radiation modulation mode for single-mode, single-frequency, and linearly polarized optical emission of the highly-coherent laser. When self-excitation conditions are met in such OEO systems, **Figure 1** generation of microwave range oscillations occurs.

In the diagram in **Figure 1**, the Laser is presented by closed into a loop the optical amplifier (OA), the optical filter (OF), which corresponds to the "traveling-wave" laser or the fiber-optical laser. The optical pump power P_p acts at the active amplifier. If the excitation conditions are met, the laser generates optical oscillations which pass from its output into MZ, then pass via two optical channels with different delays, combine together and through OF and FO acts to the light-sensitive PD area. An effective modulation by MZ is possible in microwave range only for single-mode single-frequency and linear-polarized emission of the highly-coherent laser. Quantum-Well (QW) laser diodes and the fiber-optical lasers with polarizers at their outputs are such emission sources.

The laser is the pump source for the radiofrequency network (**Figure 1b**) closed into a loop and formed by a modulator, an optical fiber, a photo-detector, an electronic amplifier, an electric filter, and a coupler.

As a result of oscillation processes, the spectra are formed with fluctuations having the various nature, but the spectral line width of radiofrequency oscillations is defined by parameters of two oscillating system: the laser and the radiofrequency oscillator.

4. Problem statement

At present, in large-dimension models of laser OEO (**Figure 1**) with the fiber-optical delay line the low phase noise level of −157 dB/Hz [5, 6] is achieved on the 10 GHz generation frequency at 1 kHz offset from a carrier.

Experimental and theoretical investigations of the power spectral density of the laser oscillator phase noise described in [16], show that reduction of the phase noise level of OEO in many respects depends on the laser phase noise level. At oscillation frequency 8 … 10 GHz at standard offsets from 1 to 10 kHz, the power spectral density of the phase noise is −120 dB/Hz … -140 dB/Hz.

Appearance on the commercial market of nano-dimension optical fibers with low losses (down to 0.001 dB per one bend, at small bend radii up to 2 … 5 mm) becomes the stimulus for improvement of OEO radiofrequency generation methods. This allows implementation of comparably small (by geometric linear maximal dimensions) fiber-optical 5μs delay lines of 10 … 30 mm.

In spite of the growth of publications devoted to OEO experimental investigations, the theoretical analysis and systematization of main mechanisms of the phase noise suppression in the low-noise laser OEO was not yet described in known literature. The laser phase noise influence on the OEO radiofrequency phase noise was not researched yet.

The purpose of the presentation is to consider the main features of OEO as a low-noise generator. This includes consideration based on the study of differential equations, the study of transients in OEO, and the calculation of phase noise. It is shown that different types of fibers with low losses at small bending radii can be used as a FOLD in OEO.

Following to an approach described in [16], for OEO noise analysis, we consider the system in **Figure 1**, in which two different oscillation processes are developed: laser oscillations with the generation frequency of approximately 200 THz and 10-GHz oscillations in the radiofrequency network closed into a loop. At that, the frequency multiplicity is approximately 20,000.

5. Laser in OEO

We will assume that the laser in OEO has high coherence and the spectral line width is much smaller than the average generation frequency, and the laser oscillations can be considered close to sinusoidal, and with a phase component of noise with normalized amplitude noises $m_{Lm}(t)$ and the phase noise component $\varphi_{Lm}(t)$:

$$E_L(t) = [E_{0L} + m_{Lm}(t)] \cos [2\pi\nu_{0L}t + \varphi_{0L} + \varphi_{Lm}(t)]. \tag{1}$$

Here $E_L(t)$, E_{0L}, $m_{Lm}(t)$ are normalized non-dimensional quantities, respectively: the instantaneous intensity, the EMF intensity amplitude, and the EMF amplitude noise, ν_{0L} is the average laser oscillation frequency, φ_{0L} is the initial constant phase shift, t is the current time.

In the opto-electronic oscillator system, under fulfillment of excitation conditions in the electronic part of such an oscillator, the radiofrequency oscillations $u = u_g(t)$ give rise. At that, the radiofrequency signal passes to the electric MZ input from the output of a nonlinear amplifier through the C coupler during oscillation generation. The instantaneous voltage of this signal is

$$u_g(t) = [U_{10MZ} + m_{em}(t)] \cos [2\pi ft + \phi_{0e} + \varphi_{em}(t)], \tag{2}$$

where $U_{0.1MZ} = U_{01C}$ is the amplitude of fundamental oscillation at the electric input of the MZ modulator or at the C output, f is the oscillation radiofrequency, ϕ_{0e} is the constant phase shift, $\varphi_{em}(t)$ are electronic phase fluctuations, $m_{em}(t)$ are electronic amplitude fluctuations.

The low-noise single-mode and single-frequency quantum-dimension laser diodes or the fiber optical lasers are used as the light sources in OEO.

The laser included in the OEO structure (**Figure 1**) is formed by (closed in the loop) the nonlinear OA, the narrowband optical filter (OF), and the optical delay line. The optical oscillation frequency ν_{0L}, which is generated by the quantum-dimension laser diodes in the autonomous steady-state, can be found (under excitation condition fulfillment) on the basis of the phase balance equations solution for the steady-state optical intensity oscillations in the optical resonator and in the laser active element.

To reveal the main mechanisms of the laser noise influence on the OEO radiofrequency noise, the laser can be described by a system of semi-classical equation with the Langevin's sources of the white noise (ξ_E,ξ_P,ξ_N), relatively, for the EMF intensity E_L, a polarization of the laser active material P_n, a population difference N. We studied the laser equation system under its operation in the single-frequency single-mode regime. At that, oscillation are linear-polarized. The main assumption for utilization of semi-classical equations is that the carrier life time on the upper operation level and the time constant T_{OF} of the laser optical filter (OF) are much larger than the relaxation time of polarization T_2. At that, the equation system with the Langevin's sources for the laser can be written as:

$$\begin{cases} \dfrac{d^2E_L}{dt^2} + \dfrac{1}{T_{OF}}\dfrac{dE_L}{dt^2} + (2\pi\nu_{OF})^2 E_L = \dfrac{2\omega^2 P_n}{\varepsilon_n} + \xi_E; \\[2ex] \dfrac{d^2P_n}{dt^2} + \dfrac{1}{T_2}\dfrac{dP_n}{dt} + (2\pi\nu_{12})^2 P_n = \dfrac{p_e^2}{h} N E_L + \xi_P; \\[2ex] \dfrac{dN}{dt} = \alpha_{NO} \cdot J_{ON} - \dfrac{N}{T_1} - \dfrac{1}{h} P_n E_L + \xi_N; \end{cases} \tag{3}$$

In (3) T_2 is the polarization time constant, the excited particles at the upper energy level, T_1 is the lifetime of the excited particles at the upper energy level,

T_{0F} is the time constant of the optical resonator, p_e is the combined dipole moment, h is the Planck constant, ν_{0F} is the natural frequency of the optical resonator on the specific n-th longitudinal mode, ν_{12} is the optical frequency of the transition, J_0 is the constant pump current, $\alpha_{N0} \cdot J_0 = (N_{02} - N_{01})/(N_{02}T_1)$ is the constant pump, ε_n is the permittivity, ν_{0F} is the intrinsic optical frequency of the resonator, P_n is the polarization of the active material, $N = (N_{02} - N_{01})$ is the population difference between the excited and unexcited levels produced by the pumping.

It should be noted that Eqs. (3) are similar to well-studied equations in the oscillator theory for the double-circuit autonomous oscillator with the inertial auto-bias chain with fluctuations.

6. Compact fiber optic delay line in OEO

At first, we would like to note that in RF FODL with geometric length of the optical fiber of 1 ... 5 km, the useful volume (in which emission propagates in the regime of one transverse mode) is not more that one cubic centimeter.

The extremely small geometric dimensions and dimensions of the FODL OEO are important for its use in on-board systems of flying unmanned vehicles, since it is possible to implement effective systems for suppressing force vibrations and accelerations and to make high-precision thermal stabilization systems.

It is comparable in size to other commercial low-noise sources of microwave oscillation. **Figure 3**, and represents a diagram of the maximum sizes for various oscillators that operate in the 10 GHz frequency range: 1 - the quartz resonator (QR), 2 – the disk dielectric resonator from ceramic alloys (DR), 3 – the disk dielectric resonator from leuco-sapphire (DDLS), 4 - OEO the fiber-optical delay line (OEO RF FODL) (delay time is 10–50), 5 - the optical disk microresonator (ODR).

Figure 3, and shows that the smallest dimensions of the resonator have ODR. The dimensions of modern microresonators, taking into account optical input and output devices, lie in the range of about 10...100 cubic microns. **Figure 3b** shows a

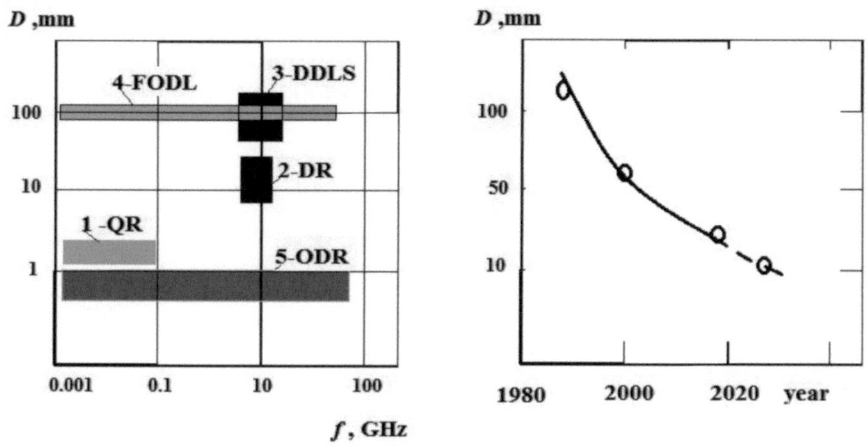

Figure 3.
Maximal dimensions of resonators and delay lines used in modern high-stable OEOs and microwave oscillators (a). Dependence of the resonator size in years (delay time is 50 μs). 1 – QR – The quartz resonator, 2 – DR – The disk dielectric resonator from ceramic alloys, 3 – DDLS – The disk dielectric resonator from leuco-sapphire, 4 - FODL – The fiber-optical delay line (delay time is 10–50 μs), 5 - ODR – The optical disk resonator. The plot of maximal overall dimensions' variations of the fiber reels in years (b).

graph of the geometric dimensions of the FODL coil of a single-mode optical fiber (the geometric length of the optical fiber is 5...10 km).

The development of microstructural optical fiber technologies with low bending losses suggests that in a few years the maximum geometric dimensions of RF FODL will be 10...50 mm. This becomes possible because microstructured nanofibers have a minimum loss of 0.001 dB per bend at a bending radius of 2...3 mm. It becomes possible to reduce the thickness of the optical shell and reduce the required volume. In this case, it is possible to apply the technology developed by the author [16] for creating fibers by the plasma method when heating the quartz fiber support tube in the temperature range from 1000° C to 19500° C.

The improvement of the technology of heating blanks of nitrogen-doped quartz glass with the help of microwave generators, the automatic movement of plasma columns along the support tubes will lead to the creation of small-sized low-noise OEO with overall dimensions of 0.5 cm3 with a delay of optical oscillation in it of 50 microseconds.

Figure 4 shows images of various RF FODL with the optical fiber length of 10 km used in OEO.

We note (**Figures 3** and **4**) that FODL geometric dimensions for the length of 10 km with the delay 50 μs is about 100x100x20 (mm^3), and dimensions of the optical disk resonator are 100x100x100 $(\mu m)^3$. The record small dimensions of FODL and optical disk resonators allow manufacturing of microwave and mm-wave oscillators in the miniature implementation with relatively high characteristics in noises and frequency tuning.

At the present stage, the geometric dimensions of the FORD AO are approximately equal to the resonator made of leucosapphyre. If we talk about using them in oscillators when generating oscillations with a frequency of 10 GHz.

Note the advantage of the linear topology of the fiber optic delay line FODL in contrast to the leuco sapphire crystal. The optical fiber in the FODL is less susceptible to extreme forces, which results in higher mechanical strength. These technical characteristics are very important, since on-board systems are subject to destructive shock effects and accelerations of several g.

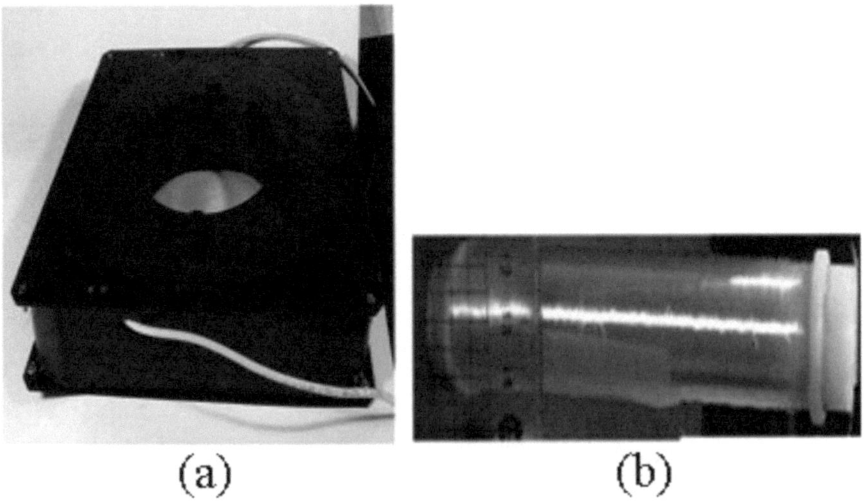

Figure 4.
Views of Fiber optic delay line (FODL) with the optical fiber length of 10 km with dimensions 100x100x20 mm^3 (a). View of FODL with optical fiber length of 0.2 km with dimensions 20x 20x100 mm^3 (b).

Note that the FODL volume consists of only 10% of optical fiber wound on a quartz cylinder. Therefore, by reducing the critical bend radius when winding the optical fiber and the cladding diameter, it is possible to potentially significantly reduce the maximum dimensions of the FODL OEO.

Moreover, in contrast to the disk microcavity, which is used in synthesizers, the nonlinear optical Kerr effect, FODL in OEO operates in a linear mode. This means that when light passes through an optical fiber, nonlinear effects and additional optical harmonics do not appear. The width of the spectral line of the laser after passing through the optical fiber in the FOLD does not change.

Figure 5a and **b** shows profiles of commercial "perforated" optical fibers with the nano-dimension structure of the light-guiding thread.

Note that microstructured fibers with extremely low optical bending losses (**Figure 5**) are used in photonic devices to generate the second optical harmonic. But for this purpose, higher power (more than 20 MW) is used at the input to the single-mode fiber. As a rule, an optical amplifier is placed after the laser or modulator.

Plots of optical losses for different types of optical fibers are shown in **Figure 5a**. From this plot, we can conclude that fibers with HALF type perforation are promising for the development of small-sized delay lines.

Figure 5a shows the dependences of optical losses for different types of optical fibers, and **Figure 5b** shows the cross-sectional profile of a microstructured optical fiber with extremely small losses at small bending radii. Analysis of the research results and optical fibers, gives the right to declare. That the HALF type optical fiber is promising for creating compact FODL [16]. Application of special or nano-dimension optical fibers (**Figure 5**) with low losses at small bend radii (1..3 mm) (0.001 dB/one bend) allows creation of miniature delay lines (1...50 μs) with overall sizes from 10 to 30 mm [17, 18].

Thus, when using microstructured optical fibers in OEO, it is possible to significantly reduce the dimensions of the fiber-optic delay line of the FODL.

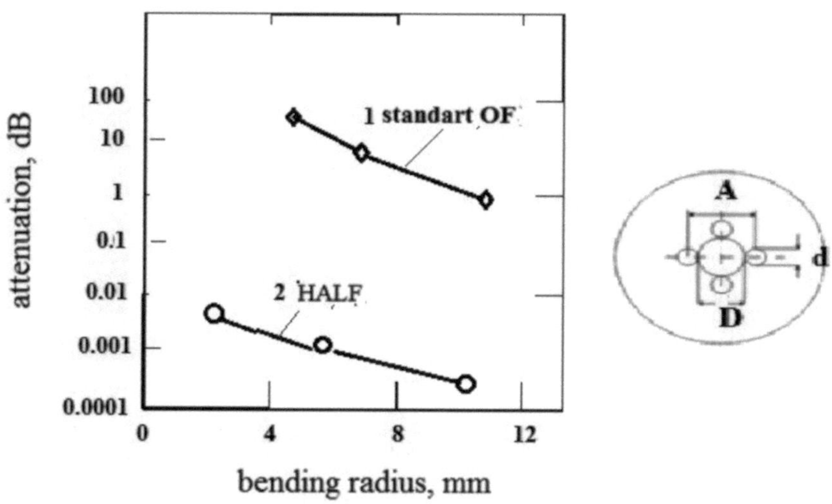

Figure 5.
a) 1)the plot of bending radius dependencies for standart optical fiber (OF) single-mode fiber G. 652 type with core diameter about 10 microns SMF-28e (corning). 2)the plot of bending radius dependencies for special microstructured optical fiber or hole-assisted light guide fiber (HALF). b)profile of commercial micro-structured "perforated" optical fiber with the nano-dimension structure of the light-guiding core.

7. OEO differential equations

To make the differential equations of a closed OEO circuit, it is necessary to keep in mind the following. The positive feedback circuit includes FOS (fiber optic system (FOS), which contains optical filters OF (**Figure 1**) optical amplifier (OA), modulators, photodetector (PD), electronic amplifier(A), electronic filter (F) and couple (C).

Taking into account the remark made, for the transfer function of the "feedback loop" K_{FB}, it is possible to write for the case of OEO DM (**Figure 1a**):

$K_{DL} = K_{FB} = \frac{i_{1L}}{E_n^2} = \frac{J_{1L}}{E_n^2}$, where $E_n = E_L$ is the normalized strength in the QWLD output, which is equal to the value of the FOS input, $i_{1L} = i_m$ is a component of the AC input voltage MZ in the OEM MZ structure and is simultaneously a component of the AC input QWLD in the OEO DM structure. We obtain the following symbolic equation for the variable component of the current:

$$J_{1L} = \frac{[\cos(\Delta\phi_{OF})]|E_n|^2(1/T_{EF})K_{OF}K_{PD}p\exp(-pT_{DL})S_{NY}(J_{1L})}{\left[p^2 + (1/T_{EF})p + (2\pi f_{0e})^2\right]}. \quad (4)$$

Taking into consideration the circuit of positive FB, we transfer to equation system in the time domain for OEO DM [16]:

$$\begin{cases} dE_{0L}^2/dt = G_0 \cdot E_{0L}^2 . N_L - E_{0L}^2/T_{0F} \\ dN_L/dt = \alpha_{N00} \cdot J_{0L} + \alpha_{N01} \cdot J_{1L} - \dfrac{N_{0L}}{T_1} - G_0 N_L E_{0L}^2, \\ d\varphi/dt = 2\pi\nu_{0P}(N_L) - 2\pi\nu_0 + \sigma_{0L} + \rho_{0L}E_{0L}^2, \\ \dfrac{d^2 J_{1L}}{dt^2} + \dfrac{1}{T_F}\dfrac{dJ_{1L}}{dt} + \left(2\pi f_{eF_0}\right)^2 J_{1L} = S_{NY}\left[E_{0L}^2 K_{FOS} K_{PD}, J_{1L}(t - T_{DL})\right] \dfrac{dJ_{1L}(t - T_{DL})}{dt^2}, \end{cases} \quad (5)$$

where the transfer function $K_{DL} = K_{FOS}K_{PD}$, K_{FOS} is the transfer function of FOS, which contains the optical fiber of two fibers of different length, K_{PD} is the transfer function of the photo-detector, which were defined in Chapter 2 in book [16].

Now we present for comparison the similar to (4.18) system from four time-equation for OEO MZ with QWLD [16]:

$$\begin{cases} dE_{0L}^2/dt = G_0 E_{0L}^2 N_L - E_{0L}^2/T_{0F}, \\ dN_L/dt = \alpha_{N00} \cdot J_{0L} - \dfrac{N_L}{T_{1L}} - G_0 N_L E_{0L}^2, \\ d\varphi/dt = 2\pi\nu_{0P}(N_L) - 2\pi\nu_0 + \sigma_{0L} + \rho_{0L}E_{0L}^2, \\ \dfrac{d^2 U}{dt^2} + \dfrac{1}{T_F}\dfrac{dU}{dt} + \left(2\pi f_{eF_0}\right)^2 U = S_{NY}\left[E_{0L}^2 K_{MZ} K_{FOS} K_{PD} \cdot U(t - T_{FOLD})\right] \dfrac{dU(t - T_{FOLD})}{dt^2}. \end{cases} \quad (6)$$

8. Dynamics of transients in OEO DM

Let us consider the transient process of the exit to the steady-state mode of the free generation of OEO DM at representation of the oscillator in **Figure 1a**. As it had been mentioned earlier, such a structure is described by the system of differential Eqs. (5). Let us describe in more detail the results of the study of the system of differential Eqs. (5) for OOO DM (**Figure 1a**).

On the base of mentioned OEO differential Eqs. (5), the analog model of OEO was constructed presented in **Figure 1a**.

Figure 5 presents the obtained solutions of system (5) and shows plots of the square of the intensity, population, and pump current, as well as phase portraits in the transient mode under the influence of a constant pump current in the form of a step pulse.

The one of difficulties at solution of (4) finding at the analog modeling is the determination of the nonlinearity of the RF nonlinear amplifier (A) in order to "compensate of the multiplicative QWLD nonlinearity".

At the same time, in the analog models, the following laser parameters were taken in solution (5) (the same as in Chapter 3 [16]): for the mesa-strip laser with the thickness of the dielectric film d = 1.2 μm: $g_0 = 10^3$, τ_D=7,2·10^{-12} s, I_{thr} = 12 mA, ε_{sh} = 0. The values of parameters of QWLD are: the life time of carriers $T_1 = \tau_{n1} =$ 0.5·10^{-9} s, the threshold level population difference is 10^{18} 1/cm^3, the life time of photons or the time constant of the optical resonator $T_{0F} = \tau_{ph} =$ 1,2·10^{-12} s, the volume of the QWLD active zone is 10^{-11} cm^{-3}.

The modes with and without delay in the OEO feedback ring were investigated. (**Figure 6**).

The pulsations of the square of the intensity and population of the laser in the simulation of the transition process OOO DM are established.These dependences are shown in **Figure 6**. The nonlinear distortions are related to the multiplicative nonlinearity of the laser. And their level depends on the value of the DC pump current of the laser.

The period of laser pulsations in transients, which is approximately 0.4 ns, depends on the level of the pump current and is determined by the carrier lifetime.

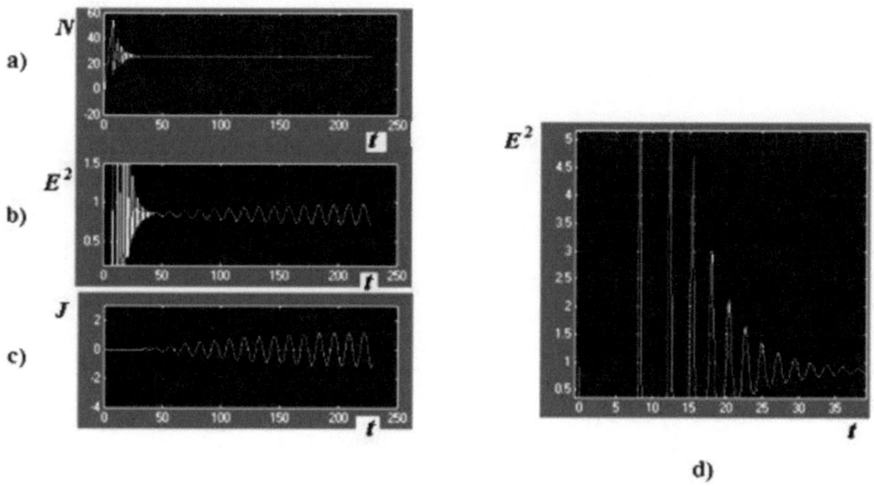

Figure 6.
The transient process in OEO and in the laser. The constant pumping level J_o = *30 mA. Activation of pumping occurs in the time moment t = 0. Time-functions of normalized values: a) the population difference N* (10^{18} *1/cm*^{3}*), b) the normalized square of strength* $(E)^2 = (E_{oL})^2$*, (1.0 point = 1 mW), c) AC component of pumping current, d) the normalized square of strength* $(E)^2 = (E_{oL})^2$*, (1 point = 1 mW) at initial part [0,50]. The transient process is presented with the exit to the limit cycle on the time diagram* E_{oL}*,* N_o*. The scale on the time axis t: 5 points = 0,1 ns. The setting time for the laser oscillations is 40 points (or 0,8 ns). The setting time of OEO RF oscillations is on the time axis t 50 ... 250 points (or 40 ns).*

The setting time of the laser oscillations is 0.8 ns (or from 0 to 40 points in **Figure 6a** and **b**). The time of setting the RF oscillations of the OEO on the time axis t is 40 ns (or from 50 to 100 points in **Figure 6**). The oscillation frequency in the steady-state mode is close to the natural frequency of the electronic filter (F) and was approximately 10 GHz.

9. OEO DM system of the laser emission

As follows from the theory of oscillations, in a transient process, a special or critical point can be a stable node, with real and negative roots p1 and p2 of the characteristic equation of the system of differential Eqs. (5). When the feedback coefficient in the OEO ring increases, the special point A becomes unstable. In this case, the characteristic roots p1 and p2 must be positive.

As shown in **Figure** 7, there is a stable limit cycle around the unstable point A. The generation is impossible if the isoclinic lines $F_1(N_L)$ and $F_2(E_L)$ are not intercepted.

The process of establishing the laser radiation oscillation ends, and then, due to the positive feedback in the OEO DM ring, there are increasing oscillations of the laser charge current, which also modulate the inverse population of the laser. This leads to subsequent oscillations of the square of the electromagnetic field strength of the laser.

If there is a single singular point A in the upper half-plane (**Figure** 7) the condition of self-excitation of the laser is fulfilled. Therefore, the excitation of the OEO DM occurs in a gentle way. This is also true for the case of an arbitrary odd numbers of nontrivial singular points. OEO DM generation may not be possible if the isoclinic lines and are not intersect (**Figure** 7). If the number of singular points is even (if there are two singular points points), the condition of self-excitation of the OEO may not be met.

Laser generation can only be excited in a "hard" way, which is initiated by a pulse from an external source.

If we consider the case of an unstable point A, the oscillatory system develops a process of oscillation growth. The nonlinearity of the electronic amplifier limits the growth of oscillations and the conditions for the existence of a limit or closed stable cycle are met (**Figure 7b**).

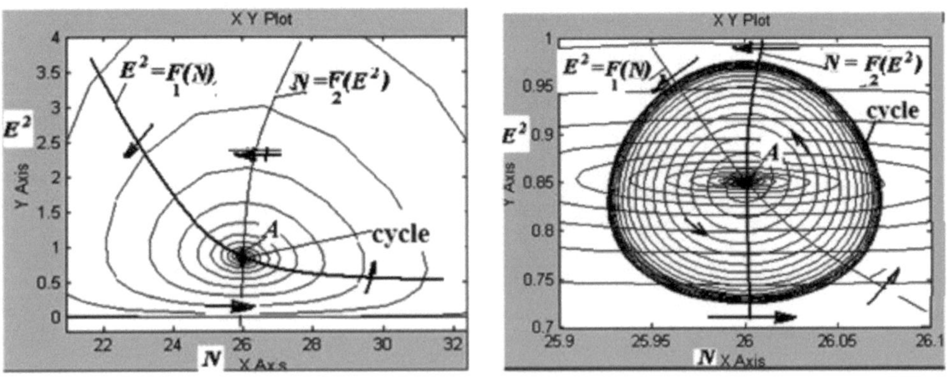

Figure 7.
Transition mode scenario in OEO DM. A phase portrait of the normalized square force is presented. a) the phase portrait of the normalized square strength $(E)^2 = (E_{oL})^2$ *and the population difference* N *and . Y-axis – The normalized square strength (or the intensity), X-axis – The population. The scale on the Y axis is 1.0 = 1* $(V/m)^2$*. The scale on the X axis is 1.0 = 1 mW.* N *(1.0 point on the scale is* 10^{18} $1/cm^3$*). b) the enlarged image of the phase portrait is shown and the transient development with the exit to the limit cycle on the time diagram* E_{oL}, N_o*. The scale on the time axis* t *is 5 points = 0,1 ns.*

If we consider the question the stability of a given oscillation cycle, then its existence is determined by the sign of the partial derivatives of the right-hand sides with respect to one variable when analyzing the characteristic equation obtained in [16], Chapter 4.

The limit cycle is stable only if the corresponding expressions for the coefficients are greater than zero [16].

When considering the hard-excited OEO DM mode, it is necessary to note more complex dynamics, and the picture of the phase plane in the transition mode becomes diverse. The number of singular points (intersection points of isoclinic lines) becomes even. Therefore, long-term generation of OEO DM oscillations in hard mode is possible only when an external generator is operating.

The transients of the oscillation tuning in the OEO DM with no lag in the positive feedback loop are shown in **Figure 8**.

Modeling has shown that strong nonlinear distortions caused by the multiplicative nonlinearity of the laser occur at a large oscillation amplitude [16] and their level is determined by the choice of the operating point or the direct pump current of the laser.

For example, a level of 1 ... 10% of the maximum possible values is performed when a constant bias current is selected at a level of 1.5 to 5.0 exceeding the threshold laser pump current. It is established that the nature of the transient process is determined by the type of non-linearity of the electronic amplifier, the

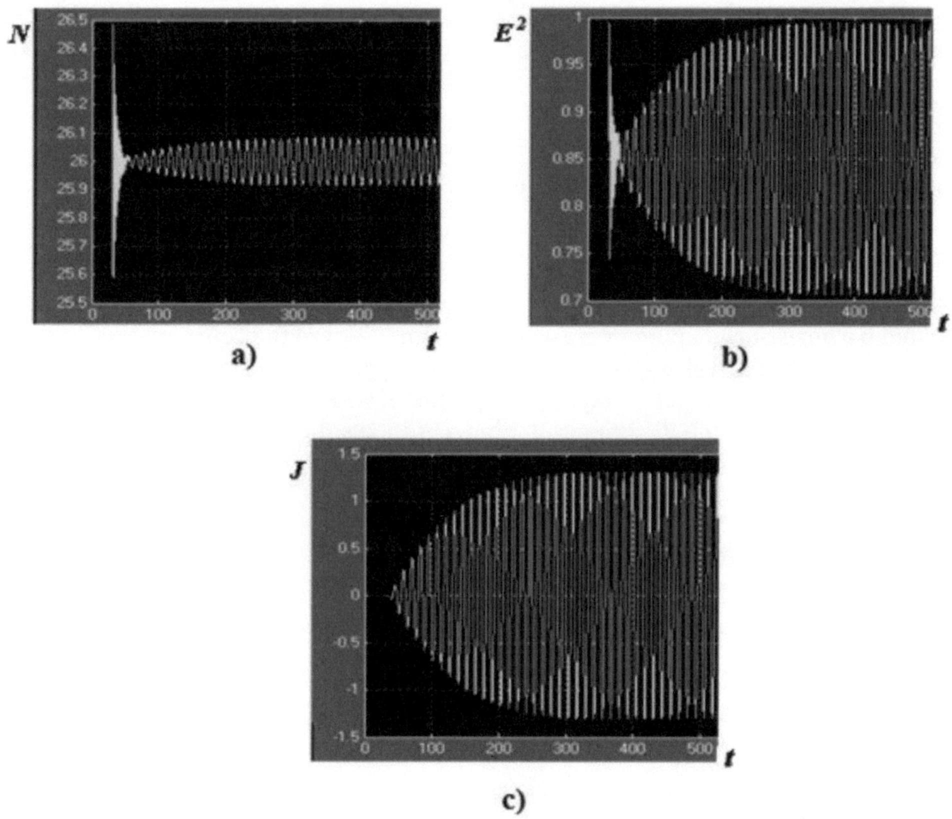

Figure 8.
Plots of function of the dependences of the square of the electromagnetic field strength of the laser in the optical channel, the population of carriers and the pump current. In abscissa axis – The normalized time, in ordinate axis – a) N – *The inversed population; b)* $(E)^2 = (E_{oL})^2$ *- intensity of laser; c), –Oscillations of OEO the electrical current of laser pumping.*

selection of the natural frequency of the electronic filter, and the delay value in the FODL. At the same time, the duration of the OEO DM oscillation transition process changes significantly.

Positive feedback is included in the DE system, taking into account the photodetection of optical radiation, selectivity in the radio frequency, and nonlinear gain on a nonlinear amplifier.

What is new in the analysis of the OEO DM operation is that the Lotka-Volterra laser differential equations for the optical field intensity, inverted population, and optical phase with positive selective feedback with a delayed argument can be reduced to a single van der Pol differential equation for the pump electric current.

From our studies of differential Eqs. (5), it follows that in OOO DM, single-frequency and two-frequency modes of relaxation oscillations are possible.

For a stable single-frequency mode of OOO DM generation, the following conditions must be met: a twofold excess of the electron filter time constant (F) over the electron relaxation time constant in the active layer of the laser.

10. Laser phase noise and OEO phase noise

Expressions for SSB PSD of the laser phase noise do not reflect the important property of the laser oscillating system: a presence of the relaxation resonance on the frequency ν_{OOL} at the offset from a carrier ν_{OL}, i.e., at $F_{OOL} = 2\pi(\nu_{OOL} - \nu_{OL})$. We can take this "resonance peak" into account at linearization of system [16] with account of the population equation. At that, the expression for SSB PSD of the laser phase noise take a form:

$$S_{PL}/P_{OL} \approx \frac{S_{SL\,\text{Im}}}{(FT_{OF})^2} + \frac{S_{LE}{D_{11}}^2 + S_{LN}{D_{22}}^2}{T_1^4\left(\left(F^2 - F_{00L}^2\right)^2 + (F\alpha_{00l})^2\right)^2} \tag{7}$$

where $F_{OOL} = (1/T_1)((T_{OF}/T_1)\alpha_0 - 1)^{1/2}$, α_0 is an excess of DC laser pumping over its threshold value, α_{00l} is a damping decrement, T_1 is the lifetime of the excited particles at the upper energy level, D_{11} and D_{22} are the constant coefficients, and S_{LE},S_{LN} are relatively, spectral densities of impacts in [15, 16] the Langevinian noise of the laser ξ_E, ξ_N, relatively. Here ξ_E is the noise of the EMF intensity E_L, ξ_N is the noise of a population difference N. **Figure 9** shows the curve 1 of SSB PSD of the

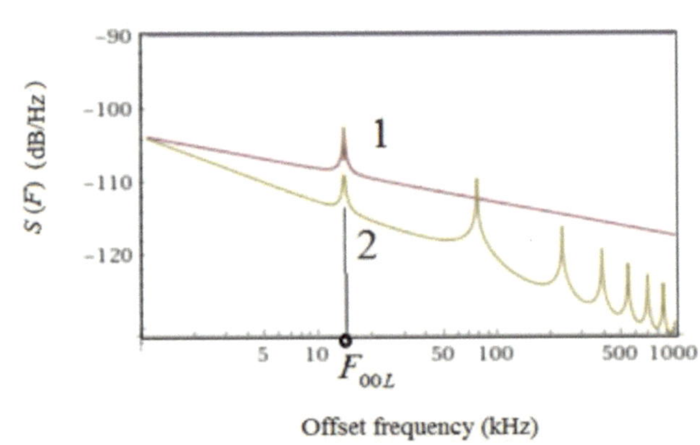

Figure 9.
Laser phase noise SSB PSD (curve 1), and OEO phase noise SSB PSD (curve 2).

laser phase noise calculated by formula (7) for $S_{SL\,\mathrm{Im}} \approx S_{LE}D_{11}{}^2 S_{LN}D_{22}{}^2 \approx -105$dB/Hz, $F_{00L} \approx 14$ kHz, the time constant of the laser resonator $T_{OF} = 10^{-7}$s.

11. OEO as the EMF correlator

We studied OEO as a correlator of two random variables ξ_1, ξ_2 with probability density at the input of the correlator $p_1(\xi_1, \xi_2)$. The random variables in the extraction of two optical harmonics [16] are phase noise $\xi_1 = \varphi_{10Lm}(t)$ and $\xi_2 = \varphi_{20Lm}(t)$ corresponding harmonics with the amplitudes A_1E_{0L} and A_2E_{0L}. The resulting phase noise of the current in the load of the PD photodetector is the result of statistical averaging.The distribution probability $p_2(\eta)$ determines the appropriate correlation function of the output process. Where $f(\eta)$ is the nonlinear characteristic of the photo-detector, $\eta_\tau = \eta(t-\tau)$ of the $\eta(t)$ process in the correlator output. At that, $p_1(\xi_1, \xi_2)$ defines the probability density $p_2(\eta)$ of the statistical process in the correlator output (**Figure 2**) of the OEO MZ (**Figure 1b**) at the *closed loop* of OEO MZ for $\tau > T_{FOS}$.

The spectral density of radio frequency OEO oscillations $S_{RFL}(F)$ is determined by the formula:

$$S_{RFL}(F) = \frac{E_{0L}^4}{2}\frac{U_{10MZ}^2 S_L}{2} K_{\Gamma PN}^2 \cdot \left[1 - \frac{A_2}{A_1}\exp\left(-\frac{2(\Delta T_M + T_{FOS})}{T_c}\right)\right], \tag{8}$$

where S_{PL}- SSB PSD of the laser phase noise (7), $K_{2\Gamma PN}^2$ is the coefficient of the noise suppression, which depends upon T_{FOS}, the laser optical power E_{0L}^2, the transfer function of FODL$|K_{FODL}|$, U_{10MZ}^2 is the square of the AC amplitude in the MZ electrical input. When considering OEO as a correlator of two random variables ξ_1, ξ_2 with a probability distribution density at its input $p_1(\xi_1, \xi_2)$, we can conclude from (8) that the SSB PSD of the laser phase noise is significantly determined by the ratio of the delay time in FOS and the laser coherence time, and it significantly depends on the ratio harmonics amplitudes A_2/A_1.

12. OEO phase noise

From Eqs. (7) with account for nonlinear characteristic of the amplifier A (**Figure 1a**) as a cubic polynomial $i_A(u) = \alpha_{e0}u - \beta_{e0}u^3$ (where u is the instantaneous voltage at the amplifier input, and the average slope of this characteristics is $\sigma_U = \sigma_{e00} - (3/4)\beta_{e00}P_{0G}$) and we can obtain through laser and delay line parameters the power of the Opto-Electronic oscillator radiofrequency generation P_{0G}:

$$P_{0G} = \frac{\alpha_{e00}}{\beta_{e00}}\left(1 - \frac{1}{P_{0L}|K_{FOLD}|\alpha_{e00}\beta_{e00}}\right). \tag{9}$$

We introduce the designation: $Y_{00}/P_{0L} = y_M[1 + FT_{EF}]/|K_{FODL}|$,

where y_M is the input normalized conductivity of the MZ modulator. Similarly to (7) for laser PSD, we obtain from the general symbolic Equations [16] the equation for SSB PSD S_Ψ of the OEO phase noise.

SSB PSD S_Ψ reduced to the radiofrequency oscillation power P_{0G} is determined by expression derived in [14–16] according to the Evtianov-Kuleshov approach.

The $K^2_{\Gamma PN2}$ coefficient depends on the delay time in the optical fiber and on the laser optical power and it is equal

$$K^2_{\Gamma PN} = \frac{\left\{(Y_{00}/P_{OL})\left[\sqrt{2}\sin\left(\pi/4 - FT_{FOS}\right)\right] - \sigma_U\right\}^2}{\left\{\left[(Y_{00}/P_{OL})\right]^2 - (Y_{00}/P_{OL})\cdot(1+\sigma_U)\cos\left[FT_{FOS}\right] + \sigma_U\right\}^2}. \tag{10}$$

where y_M is the input normalized conductivity of the MZ modulator. Then the function of OEO phase noise PSD can be represented [14–16] as

$$S(F) = \frac{S_\Psi}{P_{OG}} = \frac{K^2_{\Gamma PN} C_A h\nu N_{sp}}{P_{OG}}, \tag{11}$$

where C_A - the constant coefficient, N_{sp} is a number of spontaneous photons received by PD.

Plots of (10) are shown in **Figure 9**. which are limited functions of OEO of the phase noise PSD with account of small noises of PD and the A amplifier, at laser phase noise for the offset frequency 1 kHz equaled to about −120 dB/Hz, at laser power 30 mW, the delay of $T_{BC} = 5 \cdot 10^{-6}$s (the OF length is 1000 m), $\sigma_U = 1$. We see that the first peak is defined by the laser phase noise PSD, and average suppression of the phase noise for 50 kHz offset is more, that −10 dB/Hz.

It should be noted that at the optical fiber length of 2 km the uniform suppression of the laser phase noise is achieved in the offset range 1 ... 50 kHz.

Calculation of the phase noise suppression factor $K_2(F)$ suppression factor according to (9) is presented in **Figure 10** $\sigma_U = 1\sigma_U = 1 : T_{FOS}/T_F = 1,\ P_{OL}|K_{FODL}| = 2$ (curve 1); $T_{FOS}/T_F = 10,\ P_{OL}|K_{FODL}| = 2$ (curve 2); $T_{FOS}/T_F = 10,\ P_{OL}|K_{FODL}| = 4$ (3 curve). It can be seen that increase of delay time from $T_{FOS}/T_F = 1$ (curve 1) to $T_{FOS}/T_F = 10$ (curve 2) results in reduction of K_2 factor more than 10 times in the rated offset frequency $F \cdot T_F$ range 0.05 ... 0.5.

It is shown that at OF length, the further reduction of the OEO phase noise is possible using the PLL (phase-locked loop) system. Calculation results are well-agreed with experimental dependences of OEO phase noise PSD, which can be found in [14–16]. Here, we should remind that first publications on research of frequency stability in OEO with the help of FOLD were fulfilled in 1987–1989 at

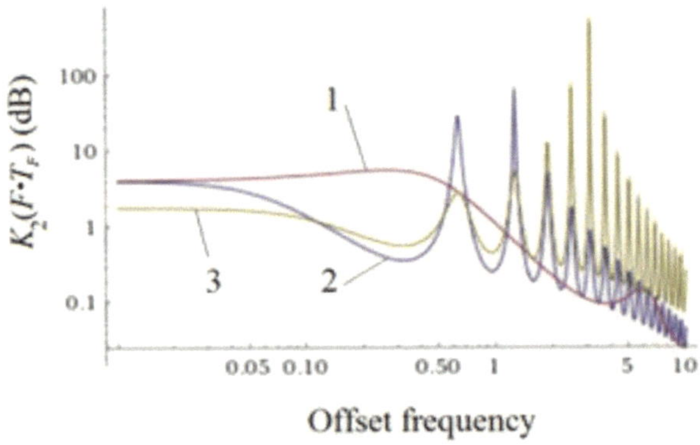

Figure 10.
The phase noise suppression factor $K_2 = K^2_{\Gamma PN}$ (9) versus the rated offset frequency$F \cdot T_F$: $T_{FOS}/T_F = 1,\ P_{OL}|K_{FODL}| = 2$, (curve 1); $T_{FOS}/T_F = 10,\ P_{OL}|K_{FODL}| = 2$ (curve 2); $T_{FOS}/T_F = 10,\ P_{OL}|K_{FODL}| = 4$ (3 curve).

Radio Transmitter Dept. of Moscow Power Engineering Institute (now NRU MPEI) while the OEO circuit was offered in [12, 13].

13. Experimental investigations

Experimental researches were devoted for several experimental OEO of microwave range with various pumping laser diodes, which emit at wavelengths of 1310 nm or 1550 nm. The maximal output power of optical emission for used laser diodes formed about 10 ... 20 mW. **Figure 11** shows the photo picture of one piece assembled on the base of the circuit in **Figure 1b**. As the photo-detector, we applied the PD on the base of InGaAs. The radiofrequency filter represented the dielectric resonator of microwave range with the loaded Q-factor of 1000. This resonator was made on ceramics and had a natural frequency 8.2 GHz. This breadboard model used the wideband (up to 12 GHz) modulation of laser emission, which was performed by the Mach-Zehnder modulator from Hitachi Co. The single-mode light guiders with lengths from 60 m to 4640 m were used for experiments. The stable generation of single-frequency oscillation at frequency close to 8.2 GHz was observed in OEO system for various OF lengths.

The delay of OEO signal was performed with the help of additional fiber-optical light guider with the 10 km-length and the additional photo-diode. The phase noise level at usage of different lasers formed the value −100 ... −127 dB/Hz, for offsets 1 ... 10 kHz from the microwave sub-carrier frequency under generation and it depends on the spectral line width of laser emission.

Essential reduction of the phase noise by 15 dB was observed in OEO using the differential delay line on the base of two optical fibers of different length. These experimental functions are well-agreed with theoretical at account of the stabilization effect at OF lengths more than 2000 m.

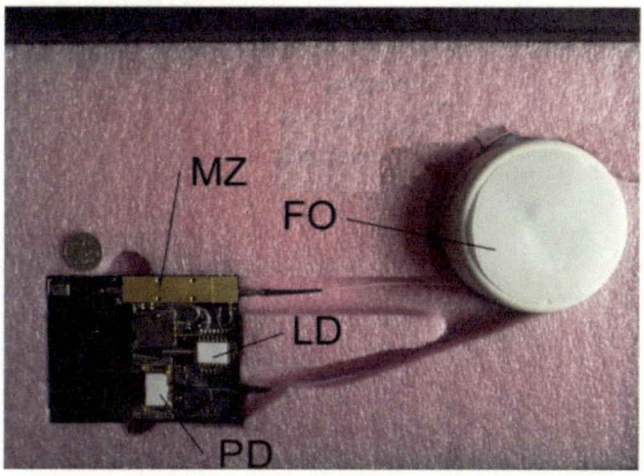

Figure 11.
General view of the experimental breadboard of low-noise laser opto-electronic oscillator of microwave range. The mean oscillation frequency is 8 ... 10 GHz.

14. Conclusion

At that, formation of the final phase RF noises of OEO is examined as the result of the convolution operation of the laser optical spectrum and the RF spectrum of the oscillation. Thus, when using microstructured optical fibers in OEO, it is possible to significantly reduce the dimensions of the fiber-optic delay line -FODL.

For a stable mode of OEO generation in the single-frequency mode, it is necessary to double the time constant of the electron filter (F) over the time constant of the relaxation of electrons in the active layer of the laser.

We have shown that the resonant curve of the electron-photon resonance of the laser has a significant influence on the formation of the power spectral density PSD of the phase noise in OEO. For stable operation of the OEO, the laser coherence time and the delay time in the optical fiber must be balanced. The use of microstructured fibers with low bending losses makes it possible to create compact fiber-optic delay lines for OEO.

Under assumption of the small and large oscillation amplitude at the modulator electrical input, we study OEO as a system in which two oscillation processes are developed on the optical frequency and in radiofrequency. The relatively simple expressions for phase noise PSD of the radiofrequency generation in optoelectronic generator in the mode with the single-side carrier with an account of the laser phase noise. The analysis fulfilled shows that under condition of predominance of laser noises being detected over v noises of the electronic amplifier and the OEO photo-detector of the filtering system.

For reduction of spurious influence of DC intensity component on the photodetector we offer to use the modulator operation mode with an offset of the optical channels “pi”.

The suppression factor of the OEO laser phase noise at optical fiber lengths from 2 to 10 km is about −8 ... -10 dB/Hz at offset of $F = 1$kHz. Utilization in OEO of the highly-coherent laser with the phase noise less than $S(F) = -100$ dB/Hz (at the same offset) is the condition of OEO small phase noises less than $S(F) = -130$ dB/Hz at the $F = 1$ kHz offset. The value of the OEO power spectral density is proportional to the spectral line width of the laser optical emission.

Acknowledgements

Authors express our thanks to PhD Yu. B. Il’in for manifested interest and participation in discussions.

Author details

Alexander Bortsov
NRU MPEI, Moscow, Russia

*Address all correspondence to: laseroeo5@gmail.com

References

[1] Aliou Ly, Vincent Auroux, Ramin Khayatzadeh «Highly Spectrally Pure 90-GHz Signal Synthesis Using a Coupled Optoelectronic Oscillator», IEEE Photonics Technology Letters , vol.30,no.14, pp.1313-1316, 2018.

[2] Dan Zhu, Tianhua Du, Shilong Pan «A Coupled Optoelectronic Oscillator with Performance Improved by Enhanced Spatial Hole Burning in an Erbium-doped Fiber», Journal of Lightwave Technology, pp.3726-3732,(2018).

[3] Zhuansun Xiaobo and etc, «Low phase noise frequency-multiplied optoelectronic oscillator using a dual-parallel Mach–Zehnder modulator», Optical Engineering 57(08),p.086101, (2018).

[4] A. Banerjeea et al., «Study of Mutual Injection-Pulling Between Two Mutually-coupled Single-loop Optoelectronic Oscillators», Optik, 11 February 2021, pp.166492-166499, (2021).

[5] A. G. Correa-Mena and etc «Performance Evaluation of an Optoelectronic Oscillator Based on a Band-Pass Microwave Photonic Filter Architecture », Radioengineering, 26 (3),pp.642-646,(2017).

[6] C. X. Li et al., «A Novel Optoelectronic Oscillator with Series-Coupled Double Recirculating Delay Lines», Advanced Materials Research, Vols. 986-987, pp. 1730-1733, (2014).

[7] Xihua Zou, Xinkai Liu, et al., « Optoelectronic Oscillators (OEOs) to Sensing, Measurement, and Detection», IEEE Journal of Quantum Electronics 52 (1):0601116, (2016).

[8] X. S. Yao and L. Maleki, «Optoelectronic microwave oscillator», J. Opt.Soc. Amer. B, Opt. Phys., vol. 13, no. 8, pp. 1725–1735, (1996).

[9] A. A. Savchenkov, A. B. Matsko, V. S. Ilchenko, and L.Maleki, " «Optical resonators with ten million finesse», Opt. Express 15, 6768-6773 (2007).

[10] C.W. Nelson et al. «Microwave optoelectronic oscillator with optical gain», IEEE, №12,v. 31,pp.152-157, (2007).

[11] J. J. McFerran, E. N. Ivanov, A. Bartels, G. Wilpers, C. W. Oates, S. A. Diddams, and Hollberg, " «Low-noise synthesis of microwave signals from an optical source»," Electron. Lett. 41, pp. 650-651 (2005).

[12] A. A. Bortsov, V. V. Grigoriants, Yu. B. Il'in « Effect of the lightguide excitation efficiency on the frequency of a self-excited oscillator with a differential fiber optic delay line» , Telecommunications and Radio Engineering, 44(8), August, 1989, pp. 137-142. (1989).

[13] V.V. Grigor'yants, Yu.B. Il'in, «Laser optical fibre heterodyne interferometer with frequency indicating of the phase shift of a light signal in an optical waveguide», Quantum and Quantum Electronics,21 (5),pp.423-427,(1989).

[14] Bortsov, A. A. , S. M. Smolskiy «Opto-Electronic Oscillator with Mach-Zehnder Modulator», Infocommunications Journal, Vol. XI, No 1, March, pp. 45-53, (2019). DOI: 10.13140/RG.2.2.20992.69126.

[15] Alexander . A. Bortsov , Sergey M. Smolskiy «Optoelectronic oscillator as the time correlator with ultralow phase noise» , Opt. Eng. 59(6) 061618 , 3 February, (2020). DOI: 10.1117/1. OE.59.6.061618.

[16] Alexander.A. Bortsov, Yuri B.Il'in, Sergey M. Smolskiy «Laser Optoelectronic Oscillators» Springer

Series in Optical Sciences, vol. 232. Springer, Cham, 522 p. 2020. –Available at: https://doi.org/10.1007/978-3-030-45700-6,(2020). DOI:10.1007/978-3-030-45700-6.

[17] Gerd Keiser, Optical Fiber Communications, McGraw Hill, 4th edition, 2008.

[18] Prajwalasimha S. N., Kamalesh V. N. Macro Bending Loss in Single Mode Optical Fibre Cable for Long Haul Optical Networks, International Journal of Emerging Technology and Advanced Engineering, Volume 4, Issue 6, June, 2014.